Colonies in Space

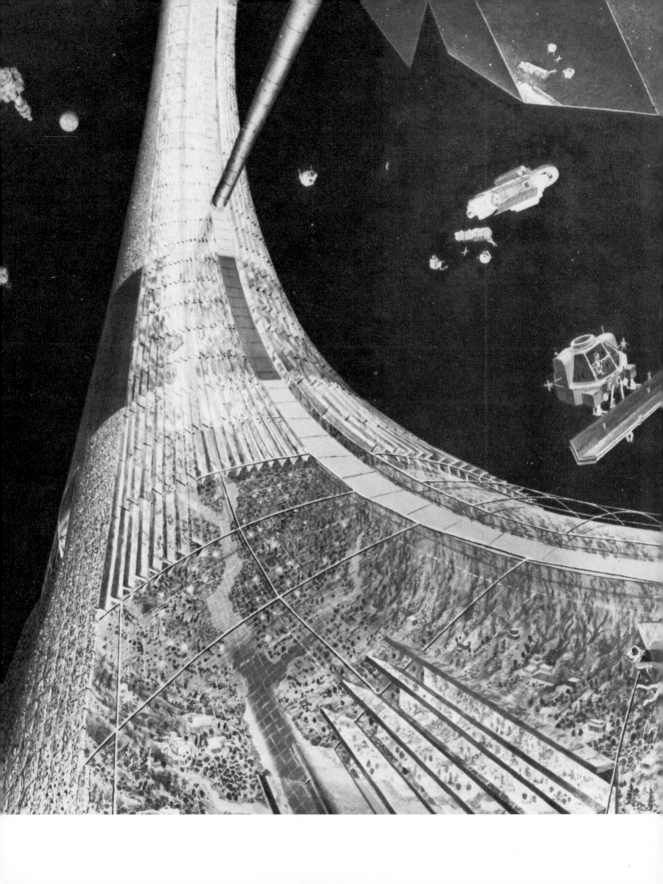

Colonies in Space

T. A. Heppenheimer

Produced by Richard C. Mesce

Original painting for jacket by Don Dixon

Stackpole Books

COLONIES IN SPACE

Copyright © 1977 by
T. A. Heppenheimer

Published by
STACKPOLE BOOKS
Cameron and Kelker Streets
P.O. Box 1831
Harrisburg, Pa. 17105

Published simultaneously
in Don Mills, Ontario, Canada
by Thomas Nelson & Sons, Ltd.

Library of Congress Cataloging in Publication Data

Heppenheimer, T A 1947–
    Colonies in space.

    Bibliography: p.
    Includes index.
    1. Space colonies. I. Title.
TL795.7.H46 1977   629.47'7   76-56187
ISBN 0-8117-0397-5

Printed in the U.S.A.

To John F. Kennedy

THE HIGH FRONTIER

Some to the rivers and some to the sea,

Some to the soil that our fathers made free,

Then on to the stars in the heavens for to see,

This is the High Frontier, this is the High Frontier.

Let the word go forth, from this day on

A new age of mankind has begun.

Hope will grow for the human race!

We're building a colony deep in space!

This is the High Frontier, this is the High Frontier.

Let us begin, for it shall take long,

Let everyone sing a freedom song.

Not for ourselves that we take this stand,

Now it's the world and the future of Man.

This is the High Frontier, this is the High Frontier.

The day will come, it's going to be,

A day that we will someday see

When all mankind is reaching out

Without a limit, without a doubt!

This is the High Frontier, this is the High Frontier.

—*John Stewart and T. A. Heppenheimer*

These lyrics were written by John Stewart and T. A. Heppenheimer to replace the original words of the song "The New Frontier," written by John Stewart of the Kingston Trio in 1962 to honor President John F. Kennedy.

CONTENTS

ACKNOWLEDGMENTS

My thanks to the people who have contributed to this book, or who have furnished unpublished material for my use. Among the latter are Mark Hopkins, J. Peter Vajk, Gerard O'Neill, Gerald Driggers, Carolyn and Keith Henson, Pat Hill, Eric Hannah, Gene Shoemaker, Eleanor "Glo" Helin, and Brian O'Leary. Particular thanks go to Don Dixon and Pat Hill, for artistic and architectural services of the highest quality; to Richard Johnson of the NASA Ames Research Center, where so much of the material of this book was developed; and to my agent, Richard Mesce.

A number of people have reviewed parts of the manuscript, or have helped me with source material, or else have expressed their continuing interest through valuable discussions. Among these are Ray Bradbury, Barbara Hubbard, Stewart Brand, Robert M. Powers, Eric Burgess, Don Wilhelm, Nancy Williamson, Rita Lauria, R. T. Jones, Eric Drexler, B. J. Long, Ed Kennedy, Don Davis, Clark Chapman, Dave Fradin, Jesco von Puttkamer, Richard Hoagland, Robert Salkeld, and Carolyn and Keith Henson.

Finally, special thanks go to Wilbur Nelson, Peter Goldreich, and Gerard O'Neill, without whom I might never have gotten involved in space colonization in the first place.

THE LIFE FORCE SPEAKS—
WE MOVE TO ANSWER

an Introduction by Ray Bradbury

Colonies in space?

The question really shouldn't be raised. For me, anyway, it is self-answering. Yes, of course. Why not? Let's move. Let's go there. Let's do the job.

But scores of millions of doubting Thomases repeat the question. Mr. Heppenheimer answers it on many levels in this book. Up front, I must make do with some sort of literary/aesthetic defense.

The old argument runs that as long as there is disease, war, poverty on Earth, we should not be taxing ourselves for the moon, Mars, or any further colonial expansions in space. Why can't we leave the asteroids alone, or Far Centauri?

As I write this, our government, run by supposedly liberal Democrats and supposedly conservative Republicans is dishing out at a lunatic pace 300 million dollars per day to finance new tanks, nuclear ships, and other exotic weaponries planned for the Day of Super Demise, when we will not only kill but simultaneously bury ourselves, with not even birds left to sing us to our rest.

If such as I speak against this, a nuclear Rain Dance ensues, promising Dooms from Russia and Darks from Peking. We continue to scare ourselves with Destruction as if it were the True Way. The Russians and Chinese stare at their atomic navels and repeat the same graveyard clichés, speak in Power tongues.

Meanwhile, the Life Force speaks to all of us. We should, we can, we *must* listen.

Why? Because, wouldn't it be terrible to wake one morning and discover, without remedy, that we were a failed experiment in our meadow-section of the Universe? Wouldn't it be awful to know that we had been given a chance, a testing, by the Cosmos, and had not delivered—had, by a loss of will and a flimsy excuse at desire, not won the day, and would soon fade into the dust—wouldn't that be a killing truth to lie abed with nights?

Mankind, en masse, could not live with such failure. We would feel God staring at us at high noon and turn away in shame.

We would not have been driven from the Garden, then, would we? We would discover that we ourselves drove us from the Garden. It was always in our power to return. Our failed imagination tossed our seed onto the infertile sands of a barren river bottom on a lost world named Earth.

So much more reason then to cultivate our gardens in space, invite ourselves back in through the gates of time and travel, and establish ourselves not just beyond the moon but beyond Mars and beyond Pluto and, finally, beyond Death.

Wherever we go, with our colonies, we will bring summer in the midst of a long winter of time. Each of us a hearth, each of us a small solar furnace digesting frosts and speaking suns.

Man the Inconstant asks after the Universe, the Constant.

Man blinks. The Universe sits and stares.

Man imprisons himself in his own bastille, not noticing he has the keys in his hands. Any time he wishes he can unlock, step forth, fly, be free.

We must become citizens of the Universe. The Universe says No to us. We in answer fire a broadside of flesh at it and cry Yes! Other worlds do not live. We will stir them with our limbs. Other parts of the Universe cannot see. We will bring a gift of eyes. Where all is silence, this thing that we call Human will speak.

Mr. Heppenheimer is keeper of the key, opener of the gate, tender of the gardens we will toss to space and inhabit with proper proportions of sorrow and joy. He offers you citizenship in the Universe.

How can you refuse?

Chapter 1

OTHER LIFE IN SPACE

Some years ago Hollywood brought out a movie titled *When Worlds Collide*. It concerned an astronomer who discovered a new planet headed on a collision course for Earth. With a little help from his friends he succeeded in building a spaceship to save a remnant of humanity after the world's destruction.

His ship got away and landed on a strange planet. The sun came out, and the people from Earth saw green fields, forest glades, and other pleasing prospects. Despite the disaster to Earth, the future of the human race was assured.

This movie—one of the last of its type made before the Space Age—expressed a hope that people have long cherished: The hope of finding new lands, new places for settlement. From the time of Columbus through the closing years of the last century this hope could find fulfillment within the limits of our own planet. In this century it has been one of the major reasons for interest in space travel.

Closely associated with this is the hope that other civilizations will be found in space, or at least other life. If we cannot visit them, perhaps we can communicate with them or invite them to visit. These ideas lie behind much current work in the field of exobiology, the search for life in space. Even eighty years ago there were proposals to create huge triangles and other geometric diagrams in the Sahara and the Siberian pine forests—any large and unused stretch of land—in hopes that the Martians would see and give us a sign in turn.

People don't want to be alone. They don't want to feel as if they are cosmic strangers without a neighbor to talk to. That is why, in a time of increasingly tight budgets for space exploration, the United States still was willing to spend $1 billion on the Viking program. Amid increasing astronomical evidence that the only life-bearing planet of the solar system is Earth, Viking's Mars landers and life-detecting instruments represented virtually our last hope of finding other life near to us in space.

In the earlier years of this century, astronomers were quite willing to accept the likelihood of life on Mars. As our knowledge advanced, scientists' views concerning the probability of life on Mars went from "quite likely" to "perhaps" to "we can't really say no" to "most likely, no," where it is today. If science can't give people what they want to believe, quite a few people are willing to believe anyway. In recent years there has been increasing popular interest in the subject of visitors from outer space.

There is something of a minor industry in this country catering to those who would like to believe that somewhere, someone has come to visit. Perhaps the leading worker in this industry is Erich von Daniken. His method is simple. Anything in archaeology that looks even a little odd is evidence that Somebody once dropped by.

For instance, there is his discussion of Easter Island. He notes that the Polynesians called it "the eye that looks toward heaven," and describes the strange stone statues scattered there. Then he says, "Aha!" and you can guess the rest.

But in Polynesian lore "heaven" is the place whence the ancestors came and has nothing to do with the sky. Easter Island is the island closest to South America. We have good reasons today (ever since Thor Heyerdahl's voyage on the raft *Kon-Tiki*) to believe that the Polynesians originally came from South America by raft. So Easter Island was most likely a religious and cultural center. Boatloads of people would come there over long stretches of ocean to help build new monuments.

When it comes to the Pyramids, von Daniken has a field day. Egyptology has been a science at least since Champollion deciphered the Rosetta Stone in 1799. In the British Museum there are samples of the heavy rope and the architects' plans used in construction. But von Daniken blithely goes ahead and writes of visitors levitating the huge blocks into place.

The theme of visitors from space actually is quite a recent one in world literature. Probably the first work written on this topic was H. G. Wells' *The War of the Worlds,* in 1895. We know what gave rise to these first stories; it was the modern era in the study of Mars, which began in the 1860s and reached a peak of popular attention after 1877. In that year Giovanni Schiaparelli observed Mars and announced that he had seen *canali*. In Italian the word means channels or grooves, but to most other people the word was interpreted as the *canals* of Mars. Schiaparelli eventually turned his attention to other matters, including fathering a line of descendants one of whom was the famous arbiter of fashion. But his reports fired the imagination of a wealthy American, Percival Lowell (of the Boston Lowells), who built himself an observatory in the pine-forested mountains and clear air of Flagstaff, Arizona.

Lowell's books *Mars as the Abode of Life* and *Mars and Its Canals* were published around the turn of the century. He described the canals as true engineering works built by a race of advanced beings facing a water shortage. The canals, he speculated, served to carry water from the polar caps. His writings fired the imagination of Edgar Rice Burroughs, who in turn inspired a whole generation of science-fiction writers, including Ray Bradbury and Robert Heinlein.

As the century progressed, other astronomers studied Mars. Some of the best observations were made high in the Pyrenées of France at the Pic du Midi observatory, where the seeing is always *magnifique*. The best observers, working in the best conditions, could not see canals. They saw instead an irregular patchwork of fine detail, which under slightly poorer conditions tended to blur into lines and curves.

By the 1950s, few professional astronomers gave much credence to the Lowellian view of a Mars transformed by inhabitants facing their own limits to growth. Still, nothing conclusive had been found to rule this out. In any case to the public Mars still was very much not only colored red but colored by the works of half a hundred science-fiction writers following the Lowellian tradition.

That was the situation on July 14, 1965. On that day, the Mariner IV spacecraft flew past Mars and returned twenty-two crude pictures. The pictures showed craters like those of the moon, an obviously lifeless body. Newspapers wrote of the "moonfaced Mars," and public belief in life on Mars suffered an abrupt blow. This affected public interest in the space program: If Mars indeed was not a new Earth, then shouldn't we spend the money on this Earth instead? In the weeks after that July 14, Lyndon Johnson committed the nation to war in Vietnam, Watts exploded in riots, and the space program was never the same again.

Some scientists still held out hope. Carl Sagan of Cornell University asked the interesting question, "Is there life on Earth?" He pointed out that there were many thousands of weather-satellite photos of Earth, each showing more detail than the best of Mariner IV's photos. Of these, only one or two showed evidence of the works of humans. This evidence, moreover, could be interpreted only if you knew beforehand exactly what you were looking at.

In 1969, Mariners VI and VII flew past Mars taking detailed photos of the south polar cap—and more craters. In 1971 Mariner IX became the first spacecraft to orbit the planet. It showed an entirely new Mars which we had completely missed on the earlier space missions. The satellite's photographs revealed immense volcanic piles, vaster than any on Earth; the largest known mountain in the solar system, Olympus Mons, 75,000 feet high; gargantuan rift valleys, ten times wider and longer than the Grand Canyon and four times as deep, and curious features resembling streambeds.

Mariner IX showed a Mars that still could hold out hope for life. Not of princesses in crystal palaces awaiting their husbands' return from the canal works, no. But apparently there once had been water there, and there once had been a thicker atmosphere. In 1976 even better photos from the Viking orbiter showed conclusively that there had been extensive

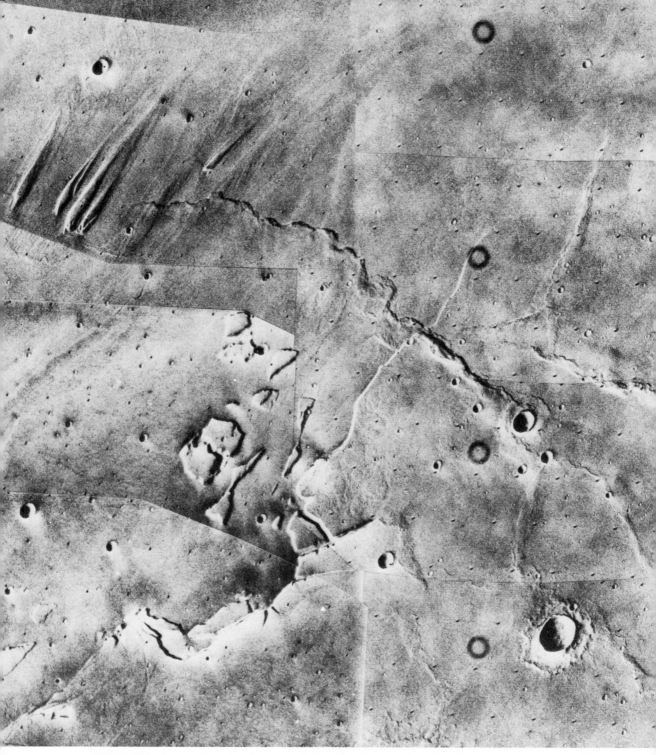

Viking photo from orbit, showing several types of characteristic Martian terrain features in Chryse Planitia. A river valley and water-shaped islands, mesas or tablelands, craters, the front edge of a lava flow, and flat desert terrain all are seen in this photo. (Courtesy Jet Propulsion Laboratory)

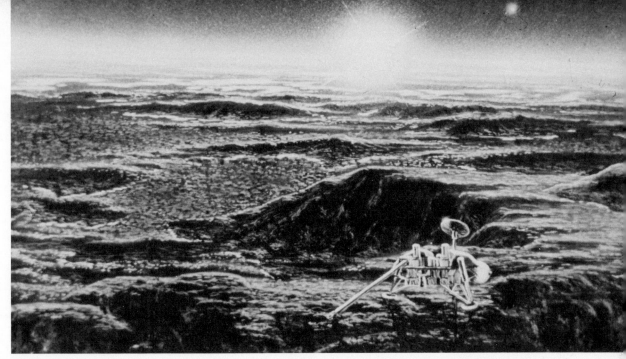

Viking lander on Mars at dawn a few days after the landing at the northernmost site. Within the lander a few ounces of Martian soil are being cultured, incubated, roasted, and analyzed for signs of life. (Painting courtesy Don Dixon)

flows of water on Mars. So perhaps there are spores, or creatures resembling bacteria, which yet might be found. The Viking landers failed to find evidence for life which could not also be interpreted in terms of chemical activity on the Martian surface. Still, Viking may not have been looking in the right places or with the right experiments, and hope for life will continue to spring eternal.

It is quite a change. In the 1890s there was wide interest in signaling somehow to the presumed Martian cities. Today the best we hope for is to find a spore or a bacterium which may stir to life in the life-detecting instruments of spacecraft such as Viking. But so compelling is the hope of life in space, so cherished is the dream, that even this discovery would be regarded as a milestone of the times.

If Mars is now to be regarded as something less than an abode of life, what of other planets? Venus was once regarded as a candidate. For over two centuries it has been known that Venus is perpetually shrouded in cloud. The clouds hide the surface below, affording considerable opportunity for speculation about hidden civilizations.

Venus is closer to the sun and receives more solar energy but its clouds reflect much of this away; so it was possible for a long time to believe that the surface would be no hotter than the Earth. In addition, Venus is nearly the same size as Earth; so there were those ready to regard it as a sister planet. A younger sister, perhaps, with dinosaurs in steamy swamps and pterodactyls in the sky.

The first solid new data to be obtained on Venus in this century demolished this picture. This data came from the first radio observations of Venus, made about 1956. By observing

VIKING LANDER 1   CAMERA 1   CE LABEL 11A
DIODE BB3   STEP SIZE 0.04   CHANNEL/MODE
AZIMUTH 132.5/252.5   ELEVATION -10(-20.22/   0.22)
OFFSET 1 GAIN 4   SCAN RATE 16K   DCS ACTIVE
DATA RATE 4000   PSA TEMP -15C(23)   DATA PATH RE
LINES TOTAL 3001   RESCAN BEGIN   0   RESCAN TOTAL
SUN AZ/EL 73.0/26.6   ANTI-SOLAR AZ/EL 13/-24
LLD/T ***/**.**.**   ERD/T 217/ 1.14.13   EVENT D/GMT
MISSING LINES   7   GAPS   4   PERCENT MISS
SOURCE TAPE/FILE DFI025/ 2   VICAR TAPE/F
SEGMENT AZ/EL/STEP SIZE 132.500/   0.220/0.04000000
ADJACENT LINE PIXELS CHANGED   227E   SAME LINE PIXELS

Viking lander panoramas of the surface of Mars. Top, Chryse Planitia (Viking 1); bottom, Utopia (Viking 2). Scientists may hope to leave no stone unturned in their search for Martian life, but they will have much work ahead of them. (Courtesy Jet Propulsion Laboratory)

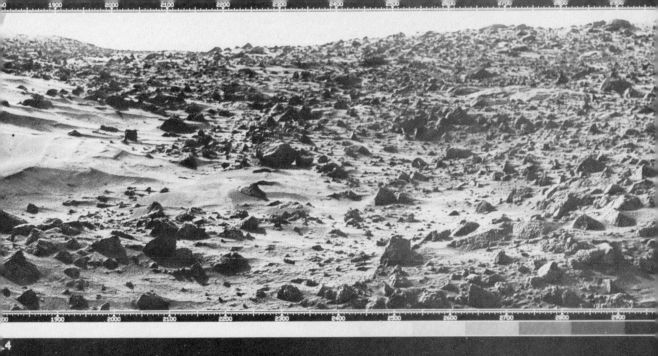

Venus with radio telescopes, it became possible to look below the clouds. These observations showed a hot Venus, at least 600°F. This was confirmed by the spacecraft Mariner II in 1962, which measured 800°.

As the 1960s went on, new data came in. Venus' atmosphere was found to be some one hundred times thicker than that of Earth. Consequently, the surface pressure would be similar to the ocean a half-mile or more down. Still, all was not lost. A trace of water vapor was discovered above the clouds, and some scientists suggested that there could be life floating high in the cooler regions of the atmosphere. This hope was dashed when it was discovered that the atmosphere contains sulfuric acid.

The picture of Venus emerging from all this is a picture of one of the more unpleasant places in the solar system. The atmosphere of Venus is, indeed, very much like the traditional picture of Hell—hot, stiflingly thick, sulfurous, gloomy under thick reddish clouds.

In the solar system the search for life increasingly focuses on places which our spacecraft haven't visited yet, or about which we still know very little. One of these places is Jupiter. Jupiter's atmosphere contains hydrogen, methane, ammonia, and water. The planet also has lightning and very turbulent weather, with vast energies playing amid the atmosphere of a world over three hundred times the mass of Earth. Ever since the work of Stanley Miller in 1953, we have known that these are potentially fruitful conditions wherein life may arise.

Artist's view of Pioneer 10 observing Jupiter in December, 1973. The bands across the planet are clouds. Io, one of Jupiter's satellites, is shown casting its shadow on the upper cloud layers. (Painting courtesy Don Dixon)

Miller prepared a mixture of those simple chemical substances and subjected it for a week to electric flashes. When he analyzed the mixture he found it had turned brown. The brown was from amino acids, the building blocks of proteins. Other investigators have since gone further. They have tested other forms of energy; shock waves appear to give particularly good results. They have changed their mixtures, heating and cooling them in various ways. These experiments have caused some of the amino acids to link up to form microspheres, small spheres resembling cells. All this has not yet produced life in a test tube but nevertheless represents important steps which nature may have followed on the road to the first cell.

Conditions on Jupiter may resemble a Miller experiment on a planet-wide scale. But further discoveries have shown that there are strong vertical winds blowing in the Jovian atmosphere. These would tend to carry any interesting molecules which form down into the lower regions of the atmosphere. Here, temperatures are high enough to destroy them, to break them up into the water, ammonia, and hydrogen from which they were formed. And from this, Jupiter looks like very much less than what Isaac Asimov hoped it would be when he wrote: "If there are seas on Jupiter, think of the fishing."

In all this, there is a common theme. Life on another planet is anticipated on the basis of the first highly incomplete data to be obtained from the planet. As more detailed data come in, the possibilities first shrink, then begin to vanish. And we are left here on Earth, apparently more and more alone in space. This could almost be a new scientific law: The probability of finding life on another planet is inverse to the amount of data we have on that planet.

Currently there is one more place in the solar system where it is said that life may exist. It's on a satellite of Saturn, Titan, 3600 miles in diameter. In 1944 the astronomer Gerard Kuiper found that Titan has an atmosphere containing methane. More recently, Donald Hunten has found that the atmosphere is nearly as thick as that of Earth. He has found it contains a good deal of hydrogen and probably nitrogen as well. These gases quite likely trap heat and raise the surface temperature to the point where Miller-type chemical reactions can proceed. Indeed, there appears to be a reddish color to the disk of Titan, seen through a telescope. This may be the color of clouds or of the surface; it may be due to the presence of complex organic molecules or even of (who knows?) life.

That is where things stand today. In 1977 and 1978, NASA will launch two spacecraft to Saturn with a prime objective of examining Titan at close quarters. If past history is any guide, the first look at these bodies will raise more questions than it will answer. The most anyone hopes for on Titan are very simple life-forms. No one is writing of Titanian cities and of princely battles over ownership of Saturn's rings.

As hope for life elsewhere in the solar system shrinks, interest in civilizations around other stars continues to grow. Here, few people doubt that if we could send spacecraft out to search, we would indeed in time find at least the spores and bacteria which are so avidly

Artist's conception of Titan, Saturn's largest moon. (Painting courtesy Don Dixon)

Artist's rendering of Saturn's rings, as seen close up. They are probably composed of millions of particles of ice-coated rock, ranging in size from pebbles to boulders, each following a separate orbit about the planet as an independent satellite. Most likely they are "leftovers" from a moon that failed to form. (Painting courtesy Don Dixon)

sought on Mars. The emphasis instead is on bigger game—advanced technical civilizations, ready and willing to communicate with us.

The world of the galaxies appears to be a friendly, congenial place for life. The astrophysicists Fred Hoyle and Freeman Dyson have shown that the processes whereby the chemical elements are formed involve a number of fortuitous coincidences which, acting together, provide this congeniality. Among the results of these coincidences is the fact that it is relatively easy to form carbon, on which life depends. And where there is life, there may in time be civilization. How many civilizations are there out there? There have been heated debates on this question with more than one scientific conference seeking an answer since about 1960. One formula for calculating the number of civilizations in space is: Multiply the number of stars formed per year, over the lifetime of the galaxy, by (1) the fraction of such stars with planets, by (2) the fraction of such planets suitably located for life to develop, by (3) the fraction of such cases in which life actually develops, by (4) the fraction of such planets on which intelligent life develops, by (5) the fraction of cases of intelligent life for which the intelligence gives rise to a technical civilization, by (6) the lifetime in years of such technical civilizations.

In this formula the factors near the beginning are items of which we do know something, but as we near the end we find that we know less and less. For instance there are about 200 billion stars in the galaxy. They have formed over some 10 billion years, the age of the galaxy. About half of these stars are in double star systems. Double star systems are very unfavorable places for planets to form because the presence of a second star will usually disrupt the quiet, even conditions needed for planets to evolve. But nearly all single stars are believed to have planets.

It seems reasonable to believe that stars similar to the sun, at least, will often enough have one or more life-bearing planets. And then things begin to get sticky.

If there is life, what is the chance of it being intelligent? It is difficult to see how there can be intelligence except in multicellular organisms. From the chauvinism of our multicellular point of view, we are prone to see the ancestral one-celled organisms as mere way stations along evolution's road to the glory that is you and I. But from the viewpoint of a one-celled creature, it may be that we and our multicellular friends are just evolution's newest creation. Life seems to have arisen on Earth some four billion years ago. For over 80 percent of the time since then, until about 600 million years ago, the most complex organisms were blue-green algae and other one-celled plants and animals.

The one-celled stage of life appears to be extraordinarily stable. For generation after generation, over billions of years, these cells grow and divide, grow and divide, grow and divide. The step up to multicellularity, with specialized organs and (most importantly) a brain, appears very difficult. By this measure, a trilobite or a primitive marine worm is far closer biologically to you and me than it is to the blue-green algae.

Did this advance to multicellularity occur relatively late on this planet? Or relatively early? Does a comparable advance occur on most planets with life? We will not truly know until we have had the chance to do detailed studies upon a dozen or so planets which not only have life but also a well-preserved fossil record. If multicellularity arises, then perhaps a few hundred million years will bring about the dawn of intelligence.

The intelligence may be like that of whales or dolphins. They lack hands and live in the water and will never develop a technical civilization. It is far from obvious that the highest and most successful forms of life must be intelligent. Life on Earth has been so successful even without intelligence, and intelligence has arisen so late in the world's history, that it could quite easily never have arisen at all. Perhaps the human race would merely be a species of smart monkeys, except for the challenges of the Ice Ages. And how many planets will experience similar upheavals in the weather? It may be that most planets experience more moderate, even climates than we have known in the Ice Ages of the past million years. On such planets life might never face the challenges which could lead to the dawn of intelligence. On planets whose climates become only slightly more erratic, on the other hand, most higher life might be wiped out as were the dinosaurs.

If civilization can arise partly through billions of years of evolution, partly through good luck, how long will it last? A culture like that of ancient China or Egypt may exist continuously for thousands of years. But such a culture could never be detected from space, lacking radio signals or similar indications of technical prowess. From the cosmic point of view, television transmission is a surer indication of civilization than is the Parthenon. We know of only one such culture, our own. We have been sending out radio signals for only a few decades.

So how many civilizations are there? The answer is little better than guesswork, but there may be quite a few million in our galaxy.

How can we communicate with them?

The first actual attempt to receive signals from them was made in the spring of 1960. This was Project Ozma. The scientists involved studied two nearby stars, Tau Ceti and Epsilon Eridani, and found nothing. We know now that Epsilon Eridani is a double star, which may mean that there are no planets there at all.

More recently Soviet astronomers using better equipment studied twelve nearby stars. They found nothing. While few such organized searches have been conducted (radio telescopes are too valuable not to be used for astronomical problems for which answers can fairly easily be found), it's a safe bet that quite a few informal searches have been carried out. It's easy to imagine that late at night, when the astronomer has finished his observing and still has some time, he turns to the night assistant and says, "Let's have a look at Eta Cassiopeiae." It is likely that if anyone does detect radio signals from a planet of another star, it will be in this way, known as "bootleg research."

The signals most easily detected would be radio beacons purposely set up and beamed toward the sun. As our radio telescopes improve, it becomes possible to study not merely

A view of our galaxy, as seen from an imaginary planet 200,000 light-years away. It is a vast whirlpool of stars—at least 200,000 million of them—100,000 light-years across. From this distance we can't even see the small yellow star that is the sun, as it carries its family of planets along on its 200 million-year orbit near the outer fringes of the Milky Way. (Painting courtesy Don Dixon)

nearby individual stars, but nearby galaxies or great star clusters in our own galaxy. The entire Andromeda galaxy, the nearest large galaxy to the Milky Way, can be observed all at one time by our largest radio telescopes. If just one planet in that entire immense array—just one civilization—has built a beacon and pointed it at the Milky Way, we might soon detect it.

So far no one has found anything of the sort. (Perhaps the Andromedans' space budget has been cut, reducing the strength of their beacon just below the detection threshold of our equipment.) But we have already tried to make our presence known in this way.

The world's largest radio telescope is at Arecibo, Puerto Rico; it is 1000 feet in diameter, built into a hollow surrounded by low hills. It could send and receive messages with an instrument of equal size located anywhere in the galaxy. Moreover, it was recently rebuilt to make it even more sensitive. When the improved Arecibo instrument was ready for use, one of the first things it did was to send a powerful beam, a radio signal, toward the Hercules globular cluster. This cluster, known to astronomers as M-3, contains thousands of stars and

The 1000-foot-diameter radio telescope at Arecibo, Puerto Rico. In November 1974, this instrument broadcast the first signal purposely produced on Earth for extraterrestrial communications, beaming it in the direction of the Hercules globular cluster. (Courtesy Arecibo Observatory)

is located some 13,000 light years from us. If there is anyone there to listen, perhaps we will get a message back—in 26,000 years or so.

It may be that radio communication is not the way to go. Perhaps most of the cosmic civilizations are millions of years in advance of us, using communication methods incomprehensible to us. They may talk only among themselves. As Carl Sagan observed in his book, *The Cosmic Connection* (Dell Books, 1975):

We are like the inhabitants of an isolated valley in New Guinea who communicate with societies in neighboring valleys by runner and drum. When asked how a very advanced society will communicate, they might guess by an extremely rapid runner or by an improbably large drum. . . . And yet, all the while, a vast international cable and radio traffic passes over them, around them, and through them. . . . We will listen for the interstellar drums, but we will miss the interstellar cables.

If this is true, there is nothing to do but to wait and see what advancing science may bring. Eighty years ago, at the height of belief in the Martian engineers, it was thought we might communicate with them by creating geometric diagrams in the Sahara, or perhaps by setting off a great charge of photographers' flash powder. Eighty years from now our current ideas may seem similarly quaint.

Is there life in space? Within the solar system, which we can reach and are now beginning to explore, the answer may be: Nothing but spores and bacteria. Perhaps the answer is: Nothing. Beyond our region of space the answer may yet be: Civilizations and cultures of greatness and magnificence untold. But we have not yet learned to detect them or to communicate with them.

As this has become apparent there has been a reaction against many of the more utopian hopes associated with space flight. Less than fifteen years ago John Kennedy could commit the nation to explore "this new ocean," with widespread hope that we were entering a new Age of Discovery. Today it is fashionable to believe that our problems can find solution only on the earth and there is nothing in space which can aid us in any way.

This is not so. If we cannot find planets fit for us to live on, or if Mars is not up to our fondest hopes—very well. We can take our own life into space. We can build colonies in space, as pleasant as we want and productive enough to markedly improve humanity's future prospects. And, we can begin to do this anytime we please.

Chapter 2

OUR LIFE IN SPACE

When the first men landed on the moon in 1969, President Nixon hailed their flight as "the greatest week since the Creation." Later that year another group of astronauts went to the moon, and on their return were feted at the White House. Also in that year the environment began to be a hot issue, and the Vietnam war became an even hotter issue. Three hundred thousand people gathered at the Washington Monument to sing "Give Peace a Chance." The Beatles recorded their next-to-last album together, "Abbey Road." Senator Edward Kennedy drove off the Chappaquiddick bridge. Vince Lombardi coached his last season in professional football.

And at Princeton University, it was Gerry O'Neill's turn to teach the big freshman physics course, Physics 103.

Dr. Gerard K. O'Neill is tall, quiet, and a modish dresser. He was a Navy non-com in the closing days of World War II, and then went to Cornell, where in 1954 he received a Ph.D. in physics. Later, at Princeton, he taught, did research in high-energy physics, and became a full professor. In 1967 he applied for one of the new astronaut positions then open, getting as far as the more detailed testing procedures in Houston before being turned down.

In 1969 he became deeply involved in the freshman physics course. He was concerned over student disenchantment with science and engineering. There were students who were very good in these areas, but many of them felt defensive about their studies because their friends were telling them that they weren't doing anything relevant. So he set up a special seminar for some of the brightest, most ambitious students. There were only six or eight of them; they met once a week for several weeks.

O'Neill's idea was to invite them to solve world problems through technology. It occurred to him that the first reasonable question to ask was, "Is the surface of a planet really the right place for an expanding technological civilization?"

Within the available time the students for the most part were only able to look things up in the library, such as the land area of the world. But one problem they were able to study was how big can you make a rotating pressurized vessel in space to hold an atmosphere and provide gravity? What would be the limitations of such habitats? Would they necessarily be small, like space stations, so that only astronauts would want to go there? O'Neill threw out this problem, and the answer came back pretty quickly. It already started to look interesting, because the answer emerged as several miles in diameter.

The next question the students considered was how much land area could you build in such habitats, using the material resources of the moon or the asteroids, which may be readily available in space? What are the limits to growth?

The first answers they came up with indicated there was more than a thousand times the land area of Earth as the potential room for expansion. They concluded that the surface of a planet was not the best place for a technical civilization. The best places looked like new, artificial bodies in space, or inside-out planets.

The classical science-fiction idea, of course, is to settle on the surface of the moon or Mars, changing the conditions there as desired. It turned out that there were several things wrong with this, however. First, the solar system doesn't really provide all that much area on the planets—a few times the surface area of Earth, at most. And in almost all cases the conditions on these planets are very hard to work with.

The idea of using a planet to provide gravity and to hold an atmosphere really represents the hard way to go about doing these things. Really tremendous amounts of material must be collected, enough to make a planet 5000 miles in diameter, before there is enough gravity to hold down an atmosphere and keep it from leaking into space. Even Mars isn't quite big enough—its atmosphere has almost entirely leaked away. Ever since Wernher von Braun published his space-station articles in *Collier's* over twenty years ago, people have been aware that a few tons of metal will suffice to build such an inside-out world, to give gravity and an atmosphere.

Also space is not an empty, hostile environment. It is a culture medium, rich in energy and in the resources needed for life. An artificial world in space gets solar energy full time, without the day-night cycles and the atmospheric absorption of a planet. Further, planets have strong gravity fields against which a spacecraft must fight. The earth's gravity is strong enough to have the same effect as a hole, 4000 miles deep, out of which we must climb. If we wish to colonize the surface of another planet, we are just climbing up a deep hole, passing through the sunshine of space—and then going down another hole.

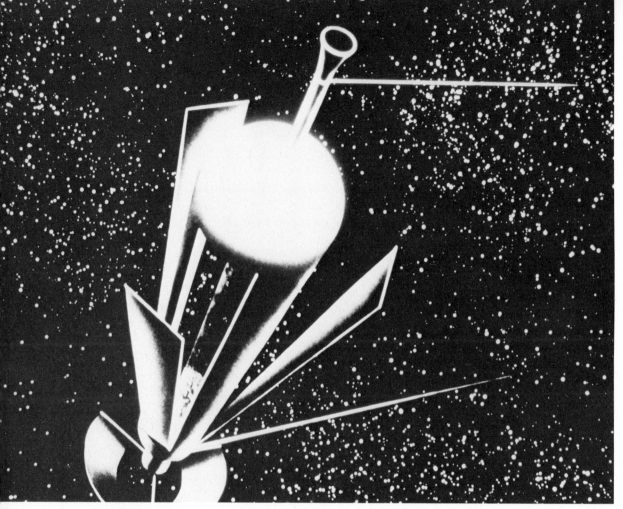

The space colony concept of Gerard O'Neill. It has the shape of a cylinder with rounded-off ends. Large mirrors, angling out from one end, reflect sunlight into the interior through window panels which constitute half the surface area of the cylinder; people live in the cylinder interior. Cables extending to the right connect this colony with another one, not shown. (Courtesy NASA)

With these interesting new ideas, Dr. O'Neill became concerned with how to make such a colony as Earthlike as possible. He certainly didn't want to invent just another design for a big space station. He wanted it to have a normal appearance inside: a comfortable atmosphere and gravity and a sun. He came up with the simple design of a cylinder whose surface would be in alternating long strips of land and window areas with large mirrors to reflect the sunlight into the colony. But this raised the problem of moving the cylinders, of shifting their position so they would always face the sun. The cylinders would be rotating to provide gravity and would act as gyroscopes, resisting any change in pointing direction. It wasn't until sometime later that the answer occurred to him: have two cylinders spinning in opposite directions and linked together. The resulting system would behave as if it were not spinning at all and could easily be made to track the sun.

Even within the first month or two, O'Neill began to believe these ideas really were practical and possibly quite important. But his work on this was quite occasional, a few minutes every week. He couldn't drop his other research and become a full-time space colonizer. Despite the pressure of his regular work O'Neill managed to write a magazine article about his space colony concept which, after much rewriting to meet the objections of various publishers, some of whom were not receptive to new ideas, appeared in *Physics Today* in 1974.

Early that year he decided he wanted to have a little conference, to try to get some people in to talk over the ideas, to see whether they were really all right or whether there might be some fatal flaw. It occurred to him that it ought to be possible to get some money to do a few things that would make such a conference a little better. So he decided to spend a few hours becoming educated in the problem of how to get money to do something entirely new.

He started calling foundations, with disappointing results. Someone suggested the Point Foundation in San Francisco, which is managed by the publishers of the *Whole Earth Catalog*.

He visited their offices while on a trip to Stanford. The Point Foundation came up with $600, which funded the first conference. It was not given without strings, however. The foundation stipulated that the money go through the offices of Princeton University rather than be given to him personally.

This turned out to be a very good thing. Because the Point Foundation's contribution was handled as an official university grant, a statement of that grant came across the desk of the university publicity office. It sent out a press release on it so that news reporters came down to that conference. One of them was Walter Sullivan of the *New York Times*.

The conference was held in May with about ten or fifteen people attending. One was Gary Feinberg, a physicist from Columbia University, who a few years earlier had proposed the existence of tachyons; particles traveling faster than light. Eric Hannah, a graduate student at Princeton, was present too. Also there was Eric Drexler, a nineteen-year-old undergraduate from Massachusetts Institute of Technology. While still in high school, he had become interested in extending Earth's civilization to the asteroids. In college he found out about O'Neill through the grapevine. There was Joe Allen, a NASA astronaut, who had also heard of the space colony through the grapevine. A couple of people from NASA Headquarters, one of them from the advanced launch vehicle group, came to the conference.

The first day they sat around a table and tried to review some of the ideas. Princeton physicist Freeman Dyson came, bringing his speculations on advanced stellar societies. He got very excited, and stayed for the second day as well. So did a few other people. The second day was open to the press, and there were a couple of local reporters along with Walter Sullivan, who wrote a news article which the editors of the *Times* put on the front page. Things began to pick up. There was a short article on the meeting in *Time* magazine. The BBC was on the phone very shortly seeking an interview; the Canadian Broadcasting

Company followed, along with various stations from New York and from the West Coast. The Associated Press did an article, and the Los Angeles *Times* requested one, too. In August there was an article in the British journal, the *New Scientist*. The *Physics Today* article came out in September, with its cover showing a painting of a colony and the words "Colonies in space."

The *Physics Today* article prompted a good deal of response from technical people who wrote in to offer objections or criticisms. These letters prompted O'Neill to work out more of the technical details, justifying a lot of things with calculations, which up to then he had just had a hunch would work out. Bernard Oliver, vice-president at Hewlett-Packard, wrote a very detailed critique. This was answered in due course not only by O'Neill but by a friend of his, Al Hibbs of the Jet Propulsion Laboratory. It turned out that there were good answers to the most important of the critics' questions.

In addition O'Neill was beginning to think about the problem of economics, of putting the colonies on a paying basis. His *Physics Today* article had mentioned some opportunities for manufacturing in space, but what was needed was some really large-scale manufacturing operation. Jesco von Puttkamer, the NASA official who controlled his funding, supplied just the idea he was looking for. Von Puttkamer suggested that O'Neill investigate the possibility of the colonies' furnishing energy to Earth. In searching for the best way this could be done, O'Neill improved upon a concept which had originally been proposed by Peter Glaser in a 1968 article in *Science*. Glaser's method of obtaining energy from space for use on Earth was to build solar power satellites on Earth and launch them from the ground by rockets. The satellites would then send electricity down to Earth via microwave beam. O'Neill's idea was to build the power satellites in space colonies, using resources from the moon. The satellites would then be moved into high Earth orbit, where they would convert sunlight into electricity. They would transmit the electricity to Earth using Glaser's method.

In early 1975 O'Neill had received a grant from NASA and was able to hire an assistant to work full-time on space colonization. He was Eric Hannah, who had acquired his Ph.D. shortly after the first Princeton conference. O'Neill also began to look ahead to two major events for 1975: a second Princeton conference and a NASA-sponsored study of key ideas. He gave several more talks, including one at Jet Propulsion Laboratory, early in February, which I attended. JPL is a hotbed of interest in space colonization, and the 300-seat Von Karman Auditorium (named for the aeronautical pioneer was founded JPL) was completely filled. I went up after his talk and met him and so was introduced into the community of space colonizers. I offered to give a paper at the next Princeton conference and eventually was invited to do this.

The second Princeton conference was sponsored by the leading technical aerospace society, the American Institute of Aeronautics and Astronautics. It was also sponsored by NASA, which provided a grant to underwrite it. At almost the last moment, the National

Science Foundation likewise came up with a grant; so the participants had their travel and lodging expenses paid.

Perhaps the most important consequence of the Princeton meeting was the creation of a community of interested specialists among the participants, thus broadening the colonization studies well beyond the work of O'Neill and his close associates.

There were a number of rather distinguished people among the conferees. Peter Glaser, the inventor of satellite power stations was there, as was Gordon Woodcock of Boeing, who had come up with a different type of design. Eric Drexler was back again, from MIT. But this time he brought along the father-confessor to his group of MIT students studying space colonization—Arthur Kantrowitz, chairman of Avco-Everett Research Labs. There was Edward Finch, former ambassador to Panama, to speak on space law. Assorted NASA officials were there to discuss what would be needed in the way of launch vehicles and how space colonization might fit into NASA plans for the future.

A great deal of useful work came out of that conference. There were key technical results involving space transportation, sources of lunar materials, and space power sources, as well as proposals for possible social and cultural organizations in a colony. Agriculture received its share of attention too. Present at the conference were Carolyn and Keith Henson, of Tucson, Arizona. They raise turkeys, rabbits, and chickens on their lot, and get their milk from pet goats. They had come to talk about farming in space, which they proposed to build around—that's right—the raising of rabbits and goats.

With the Princeton conference over, attention turned to the forthcoming NASA study, the second major event for 1975. This study was to take place during the entire summer at NASA's Ames Research Center near Stanford University, forty miles south of San Francisco. It was sponsored by the American Society for Engineering Education and represented its annual summer program in engineering systems design. This program was also sponsored by NASA to give experience in systems design to about two dozen members of university faculties chosen from around the country. One of the more noteworthy of these summer studies had been the 1971 effort led by Bernard Oliver. Cyclops, an immense array of radio telescopes to be used in seeking signals from civilizations around other stars, had been designed at this meeting.

The study started in the middle of June. Gerry O'Neill was out there for the summer to carry on his regular work in physics at Stanford. But he wound up spending most of his time with the summer study participants at Ames, whose task had been given: "Design a system for the colonization of space."

Eric Drexler again came out, this time bringing with him his entire crew of half a dozen students from MIT and Harvard. Several of them turned out to know more than most of the faculty members in the study; they did a great deal of useful work. There was Mark Hopkins, a graduate student in economics at Harvard. His economic studies helped greatly to determine the probable cost of the project ($100 billion) and the economic return from building power satellites (very high). Also there was Larry "Wink" Winkler.

Wink, as everyone called him, was particularly interested in the physiological limits to human habitation in space. He was especially concerned with the rate at which a colony should spin to provide artificial gravity. If it spun too rapidly, the colonists would suffer motion sickness. The proposal had been that the first colony should be 600 feet in diameter, rotating at 3 revolutions per minute to give normal gravity. The colony would then be a cylinder a mile long. But Wink's studies showed that there could be trouble if the spin were faster than one rpm. This meant the colony could not be a cylinder but had to be redesigned into the shape of a bicycle tire, the shape known as a torus. This is the classic shape of the space station in *2001*. It would be over a mile in diameter with people living on the inside of the "tire," 400 feet wide.

The work of the Summer Study cleared up the last major doubts as to the feasibility and practicality of space colonization. It treated in some detail such important matters as space transport of people and material, obtaining and processing metals from the moon, space agriculture, architecture and urban planning for a space community, the economics of colonization, and the provision of radiation shielding for the colony.

Toward the end of July, Gerry O'Neill went to Washington to testify before the space-science subcommittee of the House committee on science and technology. This powerful committee had recently extended its influence by taking major responsibilities for the nation's energy policies. Now it was holding hearings on possible new directions for the United States in space. Arthur C. Clarke flew from Ceylon to testify at the hearings. Gerry O'Neill gave an overall review of his work, drawing heavily on the economic studies of Mark Hopkins. Then he went off to a meeting, previously arranged, with Congressman Morris Udall.

Gerard O'Neill testifying on the subject of space colonies at hearings before the space-science subcommittee of the House committee on science and technology. (Courtesy Eleanor Crow)

Udall was a candidate for the Democratic nomination for president, and had quite a busy schedule. He agreed to spend a few minutes, however, with O'Neill. As a liberal Democrat, he had often voted against the space program. But he had heard about space colonization through the efforts of two of his constituents, Carolyn and Keith Henson. Udall was sufficiently impressed by O'Neill's presentation to write a letter to Robert Seamans, the head of the Energy Research and Development Administration, urging that the ERDA fund the design study of the space colony concept.

Near the end of August, the participants in the Summer Study held a press conference at Ames Research Center which drew a large group of reporters. Meanwhile, public interest was growing, as more magazine articles came out.

In November, the House committee on science and technology released its report based on the hearings in July. It called for a 25 percent increase in NASA's funding, or some $750 million, "to lay the foundation for advanced projects, such as moon bases and orbital colonies." The chairman of the subcommittee which had held the hearings, Don Fuqua of Florida, issued a statement: "It is hard to predict tomorrow, and although I do not have the vision to say precisely where the future will take us, I do know that our space program is only in its infancy stage."

Chapter 3

POWER FROM SPACE

The terrain around Barstow, California is a sun-blasted desert. It is a country of low hills, where scrubby brush manages somehow to stay alive, along with an occasional Joshua tree. It is monotonous country, hot, worthless for farming; among its most important uses is to provide much of the 300 miles that separate Los Angeles from Las Vegas. It is also the location of the Goldstone tracking station of the Jet Propulsion Laboratory.

In a recent series of tests there, one of Goldstone's tracking antennas was aimed not at a spacecraft millions of miles away but at a low hill a mile away. On that hill was a 100-foot tower with an array of receiver panels 25 feet high. Below the tower was a bank of seventeen lights, each about the size of an auto headlight.

Red lights flashed in the area of the antenna, 85 feet in diameter, as it swung upward upon its pedestal until it was pointing directly at the tower. Loudspeakers warned that a power beam was about to be turned on. Richard M. Dickinson, electrical engineer and project manager for the experiment, switched on the power. With the antenna zeroed in on the receiver panel a mile away, Dickinson ordered the power brought up slowly in increments of 25 kilowatts. There was no sound and at first, no glow from the lights.

Then at 50 kilowatts, the lights showed dimly under the glare of the midday desert sun. As the power was brought up to the peak of 400 kilowatts the lights brightened to full intensity, flicked on and off as the antenna was shifted slightly from left to right, or up and down.

Test of the transmission of energy via a microwave beam. In this experiment at the Goldstone tracking station, the 85-foot antenna in the foreground transmitted up to 400 kilowatts of power to the receiver panels mounted on the tower, a mile away. The microwaves were converted to direct-current electricity with an efficiency of over 82 percent and used to power the bank of lights below the tower. (Courtesy Jet Propulsion Laboratory)

This experiment represented the first high-power field trials of the use of a microwave beam to transmit electric power, a system which in years to come may provide a major new energy source—power from space. Such a system will provide the means by which electricity, generated in space from the intense sunlight available there, can be efficiently sent down to Earth.

This concept, power from space, represents one of the most important and newest ideas to appear in recent years. Arthur C. Clarke once noted that there are two types of inventions: There are achievements such as the airplane and the moon rocket, which had been anticipated centuries in advance. There are other inventions, such as radar and the laser, which had a considerable aura of surprise about them and which were predicted (if at all) only slightly in advance of their realization. Many of the latter class of inventions have grown out of progress in physics or electronics, and the transmission of power by microwave is definitely one of them. Indeed, Clarke, in his 1968 prediction of the future out there, *The Promise of Space,* completely failed to note the possibility of power sent down.

Microwave transmission of power rests upon two recent advances in electronics: a means for efficiently generating large quantities of microwave energy, and a means of directly converting this energy back to direct-current electricity.

The microwave generator is a somewhat remote descendant of the tubes first used in

35

World War II to generate radar beams. These tubes were wasteful of power, short-lived, and prone to heat up when high power levels were demanded. But the basic principle of operation has not changed over the years. Electrons are made to spiral around a magnet, to produce the microwave energy.

Over the years, these tubes were improved. They are now mass-produced in the hundreds of thousands per year for use in microwave ovens. A major advance came with the introduction of a new material for the magnets, samarium-cobalt. This not only permitted the magnets to be reduced in size by 90 percent; it also meant great improvements in efficiency, since the electrons would spiral in farther and give up more energy before striking the magnet. Another advance came through the use of platinum to supply electrons through the process called secondary emission. In this process, a small amount of electricity is made to stimulate a flow of electrons from the thin layer of platinum.

The result of these advances is a device known as the Amplitron. A single Amplitron, only a few inches in diameter, can produce up to 5000 watts of microwave power at efficiencies approaching 90 percent. If a given amount of energy is fed into a bank of Amplitrons (as direct-current electricity) only 10 percent of that energy will be rejected as waste heat. The rest will be converted into microwaves.

Microwaves are easy to form into a beam and can travel long distances with very little absorption in the atmosphere. They readily penetrate even the thickest clouds and rain to arrive at the receiving antenna, or rectenna. The heart of the rectenna is a system of small dipole antennas, similar to the rabbit ears of a TV set. Each is connected to a device called a Schottky-barrier diode. Microwaves, collected by the dipoles, are converted to direct-current electricity within the Schottky diodes with an efficiency of over 80 percent.

In the Goldstone experiments, the seventeen receiver panels on the tower mounted 4590 of these dipoles. Each one was T-shaped, and about four inches long, with the T-arms held vertically to discourage birds from roosting. The receiver panels were not designed to intercept the entire beam, but rather covered only about 10 percent of the beam area. But they demonstrated a maximum efficiency of 82.5 percent in collecting their portion of the microwave beam and converting it to electricity. Other experiments showed overall efficiencies for the total system of 54 percent, so that of the power initially fed into the Amplitrons, 54 percent was subsequently recovered from the rectenna. This is not the limit, however. Engineers such as Richard Dickinson talk confidently of achieving an overall efficiency of 60 or even 70 percent.

In an actual application, the rectenna would be several miles in diameter. The small antenna elements and Schottky diodes would be needed in the millions. They would be mounted on panels which would be set upright at a convenient angle. There would be no need for precision adjustment of these panels or of close tolerances and high accuracy in their assembly. The rectenna could be placed wherever land is cheap: in desert country, in rocky or hilly terrain, or even in the ocean. A floating rectenna out at sea, its panels bobbing up and

Receiving antenna (rectenna), which receives the power beam transmitted from a power satellite, converting its energy to direct-current electricity. Every panel contains numerous small antenna elements, each matched to a Schottky diode for power conversion. The total rectenna is some seven miles in diameter and generates enough electricity to run New York City. (Courtesy Arthur D. Little, Inc.)

down, would be a distinct possibility. The electricity garnered from the large number of individual panels would be fed through solid-state devices known as inverters to convert it from direct current to alternating current at the usual 60-cycle frequency. It would then be fed directly into the nation's power grid, being transmitted up to hundreds of miles by standard overhead power lines or by an undersea cable.

The earliest application of this technology could come about in as little as ten years. This would involve the power relay satellite, proposed by Krafft Ehricke of Rockwell International, as a means of transmitting large blocks of electric power across continents.

Ehricke's idea is based upon his observation that there is no practical way to transport electricity across oceans. Nevertheless, there are good reasons to do so. It would be possible to put nuclear power complexes in Greenland or to build nuclear power complexes on a remote island. The island of New Guinea has swift-flowing rivers which could develop enough hydroelectric power to run much of Australia—if the power could be transported across the Arafura Sea. Or the intense solar energy which falls wastefully upon tropical deserts—the Sahara, the Kalahari in Southwest Africa, the Atacama in Peru and northern Chile—could be used as a source of electricity which otherwise would find only the most limited local market. It could even become practical to generate electricity in tropic oceans, using the large difference in temperature between the sun-heated surface waters and the cold waters of the depths.

Even without these new means of generating energy, a method of transporting electricity globally would have important benefits. The most efficient way to generate electricity is to run the generating plant at full load, continuously. But electricity is not used at a constant rate. There is a peak in usage in the afternoon and early evening, and the rate of use drops off to a very low level in the early hours of the morning. So power companies must build the relatively expensive "peaking" plants which are shut on or off to cope with the changing loads. If excess power could be sent to Europe or Japan, or their nighttime power sent here, then most plants could be built as the less expensive "base-load plants," which do indeed run continuously.

The electricity produced at the earth's surface, by whatever method, then would be fed into the transmitter array, six miles square. This array would be subdivided into 1,000,000 transmitter modules each 30 feet square. Within the overall array, which would be nearly as large as Manhattan, the generated power (as much as 10 million kilowatts, which is enough to light Manhattan) would be subdivided by a network of substations and transmission lines to feed about ten kilowatts into each module. This power would feed an Amplitron, and the resulting microwave energy would be channeled into arrays of small antennas spaced every foot or so.

Each Amplitron would have attached to it a phaseshifter. The phaseshifters would control the microwave energy from the modules to ensure that the energy would be transmitted in the form of coherent waves. Without the phaseshifters, the energy would be incoherent: each module would radiate independently, and the energy would spread out uselessly. The modules would behave like a crowd of people milling around. The phaseshifters would keep the microwaves from each module properly adjusted, so that the modules would be like an army marching in step. This would produce a powerful well-focused beam, tighter than the beam of any searchlight. This beam would rise into space and reflect off the power relay satellite, which would return it to the rectenna on the ground.

The power relay satellite would be in orbit, 22,323 miles up. There it would take exactly twenty-four hours for each revolution, thus remaining always above one spot on the earth's surface, in the well-known synchronous orbit used by communications satellites. It would be a mile in diameter with a mass of 500 tons. It would serve simply as a reflector. The best reflector would not be a mirror but a mesh of thin wire, similar to ordinary window screening. This reflector would have to be maintained almost perfectly flat. It could deviate from its calculated curve by no more than a twentieth of an inch across the entire mile of its diameter. This could be done by building the reflector out of some one thousand subreflectors, each accurately flat, and each capable of being shifted in position so as to give the overall reflector its proper shape.

The satellite would be a gossamer, incredibly fragile structure, a spider web in space. It could be assembled only in orbit 200 or 300 miles up. It would then be maneuvered to its higher orbit using ion thrusters. It would need weeks or months for that trip, accelerating at very slow rates to avoid damaging the structure. Once in orbit the satellite would be a rather simple system needing to be visited only rarely, if ever.

The total system of transmitter array, satellite, and rectenna would cost some $15 billion. During a thirty-year period, the usual planning time of power companies, it would transfer so much electricity that the cost to the users could be as low as seven-tenths of a cent

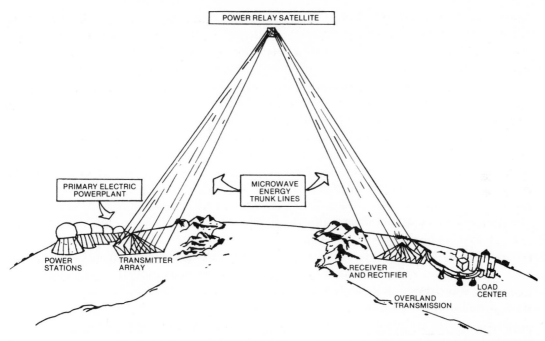

The power relay satellite system proposed by Krafft Ehricke. Power stations generate electricity, which is converted into microwaves and bounced off a relay satellite, to be collected at a rectenna thousands of miles away. (Courtesy Rockwell International Corp.)

per kilowatt-hour. It would then be no more costly than a system of overhead transmission lines a few thousand miles long—lines that could never be built across the ocean.

And it could be built within the next decade or so. The satellite would be quite large in comparison to what has been orbited, and the need for assembly in space would certainly be a new thing. But it would be only four times the weight of the Apollo lunar systems carried to orbit on the way to the moon.

The power relay satellite does not offer the promise of opening up fundamentally new major energy resources. It may permit more effective use of our existing power plants or permit the use of energy resources which until now have not been regarded as worth developing. But it would still be basically limited by the constraints which restrict our energy future. These limitations can be overcome by generating the electricity to be transmitted not on the surface of the earth, but in space.

It is nearly a century now since the first practical solar-driven engine was demonstrated at the Paris Exposition. Since then, and particularly in recent years, there has been no shortage of inventors to focus sunlight upon blackened pipes, boiling water to run a turbine or other engine. But the field of solar energy still awaits its Edison. No one yet has demonstrated the combination of low cost and high performance which would make solar power a common thing. This is particularly true for solar generation of electricity. While sunlight is free, the machinery needed to use it is not, and the equipment must necessarily spread over a large area. No solar electricity plant will ever be built as a small, compact structure with its attendant efficiencies in operation and maintenance.

What is worse, the sun does not always shine when it is wanted. In the sunniest parts of Arizona, the sun goes down long before the peak hours of electricity use are over. In the winter months not only are there fewer hours of sunlight, but the sun stays low in the sky.

In space all these problems vanish. There is at least six times as much insolation (solar influx per square foot, per year) as in the sunniest desert. There are no day-night cycles in space and no low angle of the sun in the sky. A solar collector can always be aimed directly at the sun. There are no clouds or haze. In space solar energy is available at its most powerful: constant, unfiltered, and at full strength. The problem is to use it.

The first serious proposal for a solar power satellite was made by Peter Glaser, a vice-president at Arthur D. Little, Inc., a research firm in Cambridge, Massachusetts. The original description was in an article (''Power From the Sun: Its Future'') in *Science* late in 1968. With support from NASA, he teamed up with William C. Brown of the Raytheon Company, inventor of the Amplitron, and with a group of engineers from Grumman Aerospace Corporation. The group of specialists proceeded to study the power-satellite concept in considerable detail and came up with their conclusion: it can be built, but not soon.

Glaser's powersat design is like many other satellites which have been launched because it uses solar cells to generate electricity—and there the resemblance ends. It consists of two large panels, each over three *miles* on a side, with the entire powersat seven miles long and weighing 25 million pounds. This is twenty-five times as much as the power relay satellite. The panels are not completely covered with solar cells. Instead, mirrors are used to concentrate the sunlight so that only half the areas of the panels must be covered.

The panels always face the sun. Mounted between them, free to turn to always face the earth, is the transmitting antenna, 3000 feet in diameter. Within this antenna are the Amplitrons. In the vacuum of space they need no glass tubes to enclose them. The microwave energy produced by the Amplitrons passes down hollow aluminum tubes or waveguides, escaping through slots in the direction of the earth. Phaseshifters serve to produce a tight well-focused beam.

An important result of the work of Glaser and his associates was the devising of a way to prevent the power beam from wandering off the rectenna. Their solution calls for a small portion of the microwave energy to be reflected back up to the satellite as a pilot beam, thus providing a reference signal for controlling the phaseshifters. If the power beam were to wander off, this pilot signal would be lost and the phaseshifters would fail to keep the beam properly focused. The beam would spread out, dissipating its energy harmlessly over the entire earth

Power satellite concept of Peter Glaser. Collector panels, each several miles square, are covered with mirrors which concentrate the sun's rays and with solar cells. The electricity generated by the latter is fed to the transmitting antenna (the central circular structure), 3000 feet wide. There the energy is converted to microwaves and formed into a beam. (Courtesy Arthur D. Little, Inc.)

and surrounding space. The spread-out beam would then be no more harmful than the signal from a radio station and would be as weak.

The power beam would be deliberately designed to spread out slightly on its way to the ground, to meet environmental restrictions on allowable power per square foot. Radiation leaking from a microwave oven is pulsed and can be dangerous, but power-beam radiation would be continuous, which is safer. Animals wandering into the beam, or birds flying through, would find their bodies warming up slightly. They would be in no danger of being cooked. Even in the heart of the beam, the microwaves would have less intensity than ordinary sunlight. Airplanes would find that the microwaves would bounce off their aluminum skins.

The U.S. standard for exposure to microwaves is ten watts per square foot. In the center of the rectenna there would be ten times this limit, but at the outer edges only one-tenth the limit. The main safety feature would be a chain-link fence to prevent people from wandering into the rectenna area. Microwaves are not a penetrating, ionizing form of radiation, as are X rays or the radiation from radioactive substances. They do not cause cancer or genetic mishaps, but merely warm the body.

Inevitably, some microwave energy will leak from the transmitting antenna. The leakage will be at levels far below even the strictest medical limits but will be quite sufficient to cause radio interference. It will be necessary to assign certain radio frequencies for the purpose of power transmission only. The ten-centimeter wavelength is particularly desirable for this, since it is associated with particularly high efficiencies of the Amplitron and the Schottky diode and penetrates the atmosphere well. Of course, any time there is a new allocation of radio frequencies, some people are upset. Radio astronomers will be particularly unhappy. But a space program which can build powersats can also build very large new radio telescopes deep in space and shield them from radio interference.

Unfortunately, we are no more ready to build Glaser's power satellite today than we would have been ready to build a Boeing 747 in 1935. One of the major problems lies in the solar cells. Many spacecraft, such as Skylab, have displayed impressive arrays of solar cells. But all these arrays have been assembled by hand from small individual cells, each laboriously cut and polished. If a powersat were built using such cells, its electricity would cost over a hundred times as much as its customers would be willing to pay.

The search for low-cost solar cells is under way, but it's a frustrating business. The best material is pure silicon—easily produced—but it must be available in the form of single crystals. At present, such crystals are grown slowly from molten silicon, then cut and polished by hand. A method is needed for growing the crystals rapidly in a thin ribbon or strip. Tyco Laboratories, a small firm outside Boston, has demonstrated just such a method in which ribbons of silicon are shaped with the aid of a die. Unfortunately, the silicon is so reactive that it dissolves the die. Then the ribbons which are grown are not particularly useful for power generation, because they are contaminated by material from the die.

It is quite possible that there will be breakthroughs leading to cheap solar cells, hundreds of times less costly than at present. The breakthrough may come about quite suddenly, transforming the outlook for solar cells overnight. But for now, among solar-cell scientists, there is the frustration which P. A. Berman of Caltech has expressed: "It seems almost inconceivable that such a simple thing as a solar array upon which cells are mounted in some simple, economical fashion, having no moving parts and using no exotic materials, cannot be made for a few dollars a square meter, rather than the thousands of dollars per square meter experienced in the space program."

Power from space need not rely upon such breakthroughs. It is quite possible to generate electricity by the old Paris Exposition method of focusing sunlight onto a boiler to run a turbine. Gordon R. Woodcock of the Boeing Company has pursued this approach. His powersat design uses arrays of thin plastic film, coated with aluminum, to reflect sunlight onto a hot spot which is heated to several thousand degrees. Helium gas flows through the

Power satellite concept of Gordon Woodcock and D. L. Gregory. Each of the four units has a mirror two miles wide, to reflect and concentrate sunlight in the small cavity at the apex of each set of struts. Gas turbines and generators are located at these central points. The finlike structures are radiators for waste heat. The transmitting antenna is the circle at the lower left. A solar-powered rocket is at the lower right. (Courtesy Boeing Aerospace Co.)

hot spot and drives turbines. Each turbine is hooked to a generator and the electricity produced is transmitted the same way as in Glaser's arrangement.

Woodcock's approach has the advantage of relying on the well-developed technologies of turbines, generators, and hot gases. In fact, a very similar arrangement (though without the solar mirrors) has been installed in a power plant in Oberhausen, West Germany. When such powersats are built they will be easily seen, especially on a dark night. Near midnight they will gleam more brightly than any star, as sunlight momentarily reflects from individual panels of plastic film.

It is likely to be quite a while before such powersats are in orbit, to ornament the night sky. If we wanted to build them, we would need to press our technology to the limit. There would be the classic conflict of requirements: to keep costs to the lowest level yet to use the most advanced designs. We would need structures far lighter than any yet built for use in space. The turbines would have to be guaranteed for thirty years of operation in space, yet run at temperatures hotter than any yet used for producing electricity. The nation would need to build new rockets to ferry people and equipment at one-tenth the cost of the most advanced rockets now being built. And with all these required advances, all the risky new designs, the electricity from these powersats would barely manage to be cheaper than the most expensive electricity available today. With so many new inventions needed, it is almost certain that some things would go wrong; there would be cost overruns, and the program could not pay its way.

The required launch vehicles would tax the rocket-builder's art to the limit. From the outside, they would look like enormous Mercury capsules, 167 feet in diameter and 192 feet tall. At liftoff, a launch vehicle would weigh 12,000 tons and would rise upon 20 engines, representing the thrust of 4 Saturn V moon rockets. The engines would gulp fuel (kerosene and liquid oxygen) at the rate of sixty *tons* per second, lifting a vehicle the weight of a naval cruiser. The power of ten such cruisers would be needed, to run the *pumps* of such a rocket.

To keep costs to the lowest level, the rocket would be fully reusable. All propellant tanks would be enclosed within the framework of the huge structure. The broad flat surface at the base of the rocket would give protection while re-entering the atmosphere. The rocket engines set into this base would need special protection. This would be provided by carrying water in the base. During re-entry the water would boil into steam which would flow to the outside of the base and shield the engines. There would also be a number of smaller rocket engines to fire just before touchdown and provide a safe landing.

Such a launch vehicle would resemble a ship in more than its size and weight. It would take off and land on water—on an artificial lagoon at Cape Canaveral three miles across. A creature of water, air, and space, it would never touch the land. After each flight it would be towed to a berth or drydock to be made ready for the next flight. It would carry 500 tons of cargo to orbit. Yet it would be necessary to launch such a rocket nearly every day to build one powersat a year.

One of the earliest projects would be to build in orbit, 250 miles up, an assembly facility for powersats. It would have its own electric power supply, living quarters for 100 people, communications, and other major functions. Its framework would support a system of tunnels, propellant lines, docking ports for spacecraft, and an assembly area. There would be a propellant depot and room for the loads of equipment as they arrive.

The assembly facility would serve primarily to build the thermal engine systems for the powersats. The first powersat parts to arrive would be the panels for the hot spots, or thermal absorbers. For reasons of safety there would be a manned tug to meet each load at some distance, with a pilot to guide the load to a docking port at the assembly facility. Other tugs would disconnect panels from these loads and transfer them to the appropriate locations in the assembly frame, clamping them in place. The panels would also be attached to each

Preparation for launch of a large rocket such as would be required for launch from Earth of a power satellite. The rocket is a single-stage which can fly to orbit with 500,000 pounds of payload and return. It stands 192 feet high. Preparatory to launch, it has been towed through a lock and positioned over flame deflectors; the water level has been lowered to expose its twenty rocket engines. (Courtesy Boeing Aerospace Co.)

Liftoff of the rocket. Its weight is 12,000 tons and its engines have four times the thrust of the Saturn V moon rocket. (Courtesy Boeing Aerospace Co.)

Following on-orbit operations, the rocket re-enters the atmosphere and plummets toward a large artificial lagoon on which it will splash down. (Courtesy Boeing Aerospace Co.)

Immediately prior to splashdown, rockets fire to slow the descent. (Courtesy Boeing Aerospace Co.)

The rocket-launching facilities at Cape Canaveral. The Vehicle Assembly Building is at center left; Pads 39 A and B are at bottom. These are used for winged rockets which land on the 15,000-foot airstrip. The three-mile-wide artificial lagoon is used for the landing of large single-stage rockets. In the upper right, a rectenna built upon swampland provides electricity to support the requisite large-scale production of hydrogen propellant. (Courtesy Boeing Aerospace Co.)

other. Additional modules would be attached, with elements of the turbines, generators, and heat exchangers. Collapsible frames would be mounted and extended to form a structure for the radiators. When the radiator panels were attached, the resulting relatively compact systems within the assembly frames would represent over half the total weight. Like the hull of a clipper ship, this system would be ready for its superstructure. The solar reflectors, resembling the masts and sails of a clipper, would be built up. Workers in specialized pods would attach collapsed support arms, extending them to form the structure supporting the reflectors. The reflector panels would be added last of all.

The first power produced would serve to ferry the powersat to geosynchronous orbit. Propellant tanks would be attached as well as electric-propulsion engines. In these engines the electricity would operate a number of carbon arcs to provide high temperatures. Hydrogen gas would flow through the arcs and be heated, providing several thousand pounds of thrust. The powersat would slowly accelerate, reaching its final orbit several weeks later. A few of the workers would stay aboard to make sure that it was functioning normally. They would then return to the assembly facility in a small space tug carried along.

Assembly on-orbit of elements of the power satellite shown on page 43. The central cavity structure is supported by lightweight trusses, extending downward to the two-mile-wide mirror. The U-shaped structure is a temporary framework for construction of the radiators, which are the large panels above the central cavity. On the right arm of the U, a 100-foot-long spaceship indicates scale. (Courtesy Boeing Aerospace Co.)

There is no doubt that in time these things will all be possible. It will even be economically feasible to do them. Right now, nothing of the sort is at hand. At the current rate of progress in space technology, they may become feasible sometime in the next century. They will represent a twenty-first-century solution to a twentieth-century problem: the need for new energy sources.

Fabricating powersats on the ground, then sending them up the earth's 4000-mile-deep gravity hole to space, one by one, is a needlessly expensive way of solving the energy problem because the cost of overcoming gravity is like a tariff which nature levies upon space flight.

The powersat components could be built in a space colony rather than on Earth. The colony would require a single large effort in order to be built, using resources obtained—but only in part—from Earth. Using resources from the moon, the colony can turn out powersats in space—as many as would be required. Instead of paying the "gravity tax" every time a new powersat is built, this "tax" need be paid only once—at the outset of the colonization program.

The colonization effort would begin by building in Earth orbit a "construction shack" resembling the orbital assembly facility. In addition to possessing complete assembly facilities, the construction shack would also have ore-processing facilities to extract metals from lunar rock. It would also have a large power plant delivering energy for ore-processing.

While the construction shack was being built to hold 2000 workers, a smaller crew of 100 or 200 would establish a lunar mining base. The major facilities would be a nuclear power plant, to allow operations through the lunar night, and a mass-driver. The mass-driver would accelerate payloads of lunar rock and soil, launching them to space, where they would be collected by a mass-catcher.

The construction shack would be moved to the site of the colony and the lunar work crew would begin sending up a stream of material. This material would be transported to the construction shack to be processed into aluminum, magnesium, titanium, iron, glass, and oxygen—all of which are abundant in lunar rocks. The metals in turn would serve to build the colony and to build powersats as well. A few powersat components, such as turbine blades and Amplitrons, might still be brought from Earth. But over 90 percent of the powersats' mass might be furnished by the metals obtained at the colony. Moreover, it would be easy to propel the powersats from the colony site into geosynchronous orbit. This maneuver would call for a velocity change of less than half that needed to transport a powersat from low Earth orbit to geosynchronous orbit.

There would be other benefits as well. Building powersats at a space colony means that it would not be necessary to strain the limits of technology. Instead of requiring the lightest possible structures and the hottest turbine operating temperatures, it would be perfectly feasible to use the types of designs which are available at present. When Gerry O'Neill was discussing this point with Gordon Woodcock, he asked what kind of power plant weights Woodcock was looking for. Woodcock replied, "Five kilograms per kilowatt. We can reach that by 1990." O'Neill replied, "With my system, all we need is ten kilograms per kilowatt. When could we have that?" Woodcock answered, "Why, we could have that today!"

Chapter 4

HOPE FOR THE FUTURE

Since the mid 1960s a number of premises about the basic economic structure of the world have become widely accepted. Among the more important of these ideas are:

(1) That every major human activity must be confined to the earth's surface

(2) That humanity's resources of energy and raw materials are merely those of planet Earth

(3) That these resources are limited essentially to those now known to exist, so that the world's economics must henceforth be based upon redistribution of resources rather than upon development of major new resources

Much of the rationale for this point of view lies in a small book which was one of the publishing phenomena of recent years—*The Limits to Growth* by Donella and Dennis Meadows, Jorgen Randers, and William Behrens. Like much that is characteristic of our age, the influence of this book may lie less in the cogency of its argument than in the public-relations aura in which it was born.

In a sense the book began in 1968 in the mind of an international industrialist, Aurelio Peccei. He is an executive of both Fiat and Olivetti, the Italian corporations. He was also one of the founders of the Club of Rome, a by-invitation-only association of prominent scientists, businessmen, and politicians which *Science* describes as "bearing an uncanny resemblance to Jules Verne's fictional Gun Club of Baltimore," organizers of the famous lunar flight. It was Peccei who organized a scarcely less ambitious effort, The Project on the Predicament of Mankind. This predicament, the Club of Rome believed, lay in the unquestionable certainty of the three premises mentioned, which they felt meant ultimate doom for mankind unless humanity would repent of its sinful, growth-seeking ways.

For two years members of the Club of Rome carried their message from Moscow to Washington to Stockholm to Rio, seeking to warn world leaders of the coming apocalypse. They were treated courteously, but found their words alone would not turn mankind's course. Something stronger was needed: a computerized study which would prove conclusively that what they were saying was true.

After several months of searching, they settled on Jay Forrester of MIT as the man to do the study. Forrester has had a distinguished career both in designing computers and in using them to study difficult economic problems. He agreed to do the job at a Club of Rome meeting in Berne, Switzerland on June 29, 1970, "that momentous date when it all began," in the words of one Club member. He conceived the main features of his mathematical model of world growth trends while aboard the plane returning to New York. Shortly afterward Eduard Pestel, another Club member and a director of the Volkswagen Foundation, arranged a $250,000 grant from his foundation to support the work.

The model used involved five quantities: population, pollution, food production, industrialization, and consumption of resources. Each of these was taken as representing activities over the entire earth. Forrester and his associates defined about 100 relationships among the variables (such as the relation between industrialization and the birth rate) and described these relations by equations. They proceeded to run the model, or set of equations, on a computer, making various assumptions about policies which might be followed over the entire world.

The results were what Peccei and his associates were hoping for: Unless world trends in population growth and industrialization are checked and pollution severely curbed, civilization faces a catastrophic collapse within 100 years, and perhaps within 50.

The Limits to Growth was actually written, for the most part, by Donella Meadows, the biophysicist wife of systems analyst Dennis Meadows. Late in 1971, under the encouragement of Peccei, the Meadows signed over the rights to the book to a Washington think tank, Potomac Associates.

To William Watts, Potomac Associates president, the book was (as a later press release described it) an "intellectual bombshell." With the aid of the Woodrow Wilson International Center, he arranged a symposium on the book to be held at the Smithsonian Institution, with financial aid from Xerox. To get proper publicity Potomac Associates hired a local public relations firm. They turned out press releases and background material, released their material late in February of 1972, and hit the jackpot. The *New York Times,* the *Washington Post,* and other major papers splashed the story across their Sunday editions. Columnists soon picked up the story, describing the book's "shattering insights" into the calamities facing mankind.

Suddenly the symposium at the Smithsonian was an event of major proportions. There

51

was a flood of phone calls from ambassadors, Cabinet officials, congressmen, and scientists, all clamoring to attend. The meeting was held on Thursday, March 2. That morning, the first copies of the book were released for public sale. Hardly anyone at the symposium had a chance to study the book or to review it critically.

The symposium opened with a blaze of kleig lights and TV cameras. Hardly anyone challenged the conclusions of the work (how could they?) but the ambassadors and government officials present discussed at length the work's implications for social policy. Secretary of Health, Education and Welfare Elliot Richardson pronounced the work "too thoughtful, too thorough, too significant to ignore."

Not all were caught up in the excitement. At noontime that day, a group of reporters held an impromptu news conference. They asked Meadows why he had rushed to publish a popular book before first submitting the work to the criticism of professional economists in technical journals. His reply: "Journals take so long. You're talking about delays in publication of twelve months on up." He said that he would in due course publish a technical report, describing the study "equation by equation." Carroll Wilson, a fellow scientist at MIT and a member of the Club of Rome, said that "so few will read the technical report and so many will read the book that it doesn't really matter."

As the symposium went on, critics of *Limits* began to find their voice. One warned, "The masses will look at these diagrams and believe them, but I feel it's dangerous to speak of projections so far ahead. If we feed the decision makers half-baked conclusions we can do great harm." Another critic said that "this is not a decision-making model," urging substantial refinements. But such views were in the minority. Eduard Pestel remarked that "policy decisions can now be derived from what has been worked out. There's no need to wait to start action." Edward P. Morgan of ABC Radio said that negative reaction to the book would come mostly from "reactionaries and older folk," and said, "It's up to us, the news media, to mount a basic education campaign here." Aurelio Peccei explained that the book was "a tool of communication to move men on the planet out of their ingrained habits." To make sure that the tool would not rust from lack of use, he announced his intention to translate the book into half a dozen languages and to send it without charge to 12,000 world leaders.

At MIT the enthusiasm was considerably less than universal. One associate of both Meadows and Forrester, a senior scientist who asked not to be named, put it this way: "I happen to like Dennis Meadows. He's a nice fellow and very bright, if he doesn't go off the deep end. I find their work fascinating, but I'm not about to tell a congressman to base his career on it. . . . What they're doing is providing simple-minded answers for simple-minded people who are scared to death. That's a dangerous thing. There's a sense of naivete here too . . . it's not that *they* want publicity or a grant, but they want to save the world. This messianic impulse is what disturbs me."

Aside from the public-relations effects and the messianism of the Club of Rome, just what substance is there to the work? The work basically represents an attempt to improve decision-making processes by substituting presumably exact computer methodologies for the fuzzy, often erroneous verbal methods in use. The authors of the book would argue thus: People can easily understand relationships between individual components of the world system, such as the relation between pollution and the quality of life. But when there are a great many of these relationships, human minds are not very good at deducing the effects on the entire world of all the relationships. However, it is easy to use a computer to study the effects of any number of relationships. Human intuition is inferior to the workings of the computer in seeking policy decisions.

Computer users have an expression, GIGO, or Garbage In, Garbage Out. That is, a computer's results are no better than the information fed in, or the mathematical model used to process the information. While an impressive set of results may be produced using a particular model, the results are worthless if, by using an equally plausible but different model, results are found which are completely different.

In the case of the world's future economy, there are two points of view which may be regarded as ideological poles. These are the Malthusian view and the technological-optimist view.

The Malthusian view is a latter-day version of the outlook of Thomas Malthus, an advocate of limited growth in the 1700s. He argued that population would grow exponentially while available resources would grow only linearly, so that ultimately there would be a catastrophe. The modern advocates of this view argue that the earth has only a finite amount of resources, such as agricultural land. Further, they argue that anything which makes life better, such as rising living standards or better nutrition, acts to promote population growth. New technology can only temporarily alleviate shortages, since population growth must inevitably overtake any increase in production.

The technological-optimist view holds that there are no foreseeable limits on the production of goods. If there is a shortage in a particular resource, it can be eliminated by development of alternate resources, or substitutes. Increases in the standard of living lead to lower birth rates. Eventually, there will be zero population growth from the lowering of birth rates, and technology will then provide a continually rising living standard.

The model used in *The Limits to Growth* is basically Malthusian. The quantity of natural resources is assumed to be fixed, and the productivity of industry is thought of as decreasing with time. In agriculture, it is assumed that nations can increase productivity only by increasing capital investment. The pollution output, in proportion to the material standard of living, is regarded as irreducible. The birth rate increases strongly with increasing food per capita, decreases strongly with increasing pollution and crowding, and decreases only slightly with increasing standard of living.

The future, as predicted by *The Limits to Growth,* is keyed to the steady depletion of a fixed supply of natural resources. As the supply diminishes over the next century or so, the population continues to increase. The early decades of the next century bring about a crisis in industry, as these trends continue. Capital investment in industry peaks out and declines, as does capital investment in agriculture. The standard of living then begins to decline rapidly, and the population also begins to fall as the world quality of life deteriorates.

But do we find similar results if we adopt the viewpoint of a technological optimist? This question has been studied by Robert Boyd of the University of California. He started with the basic Forrester model, with its five variables. To these, he added a sixth: technology. He wrote equations to express the assumptions that increased capital investment would speed technological growth; that increasing technology could reduce pollution, or could be used to improve the quality of life, or increase agricultural output. He also assumed that technology could reduce the amount of natural resources needed to maintain a given living standard. The birth rate, in Boyd's model, would decrease strongly with increasing living standards and would not increase with increasing availability of food.

The results of this model were exactly what a technological optimist would expect. In these results, technology is seen to increase the living standard, which in turn drives down the birth rate. The population in time levels off. The quality of life (an index derived from such factors as surplus food per capita, disposable income, crowding, and pollution), approximately constant from 1950 to 2050, then begins to rise. The standard of living rises slowly to about the year 2020, then rises rapidly.

Forrester has argued that the world system often will show behavior contrary to what one would expect from intuition. He introduces birth control into his model by reducing the birth rate—and finds that the eventual catastrophe is even worse. In Boyd's model birth control greatly improves the situation, since the quality of life then is found to increase almost continually.

Boyd's conclusions, which were published in the August 11, 1972 issue of *Science* but which received far less attention than they deserved, show that *Limits* has its own limits. The computer models used appear as powerful methods for testing assumptions. But they do not infallibly foretell the future. A different set of assumptions does give entirely different, and far more hopeful, results.

One area where the question of limits to growth impinges upon public policy is the problem of energy. In this country there has been much wailing and gnashing of teeth over the energy crisis. This crisis is not due to actual shortages of energy supplies. What is happening is that the United States is in the early stages of a transition from reliance upon petroleum as an energy source to reliance upon alternate sources of energy. A rather similar transition occurred early in this century as petroleum and natural gas supplanted coal. An even earlier transition, in the last century, put the New Bedford whalers out of business when kerosene for lighting replaced whale oil in the nation's lamps.

While environmental concerns make some sources of energy more costly, these concerns are not responsible for the underlying problem of energy. The Alaska pipeline was stalled by five years of lawsuits, many of them inspired by concern lest the builders disturb the Alaskan elk and caribou. An act of Congress swept aside these objections. But the oil from that source, however valuable and necessary, will be gone before the next century is very old. Similarly, the world's uranium will suffice for only a few decades of large-scale operation of conventional nuclear plants. Then the problems begin.

We have enough coal to last a thousand years. But much of it cannot be recovered without ripping up vast stretches of western lands in states like Utah and Montana which do not now propose to become "energy colonies" of Los Angeles. We have shale oil in abundance, but it is costly to transport and cannot be processed on the spot (in Wyoming and Colorado) without the water which that region conspicuously lacks. Moreover, the only long-term source of nuclear energy which can be built with present technology is the breeder reactor. The breeder uses plutonium, the stuff of nuclear bombs. An energy economy based upon the breeder would involve many tons of the stuff, to be transported to and from reprocessing plants. Ten pounds, in the hands of a hijacker or terrorist, would suffice to build a small but potent bomb.

The Energy Research and Development Administration, ERDA, is the federal agency charged with solving these problems. There is little doubt they can be overcome, in time. Strip-mined lands can be restored, at a price. In Germany, the Rheinbraun Coal Company has routinely done this, even building new towns on the restored land. Oil from shale may be obtained by heating the rock with fires lighted underground. Breeder reactors and their fuel-processing plants may be sited together in huge nuclear parks, built perhaps on artificial islands, to provide security against plutonium theft. Still, such solutions will be costly. Following the Arab oil embargo, President Nixon proposed "Project Independence" to meet America's energy needs. The price tag: $600 billion to $2 trillion.

And at this point the idea of solar power satellites, built in a space colony, becomes quite appealing.

Power satellites, of course, tap the inexhaustible energy of the sun. Such inexhaustible energy sources are in short supply. We will continue to get something out of hydroelectric power (though all the good rivers have been dammed) and we will get something from the winds, something from the tides. There will be other renewable sources such as firewood or household garbage. These sources will be appreciated but will not solve the problem. Fusion power is nearly as inexhaustible as sunlight, but it has not yet even been demonstrated in a laboratory and no one can say when this will be done. Until this happens and until we learn precisely what fusion power will entail, we cannot count on it either.

The power satellites' rectennas are very kind to the environment. They emit no fumes or smoke, produce very little heat, foul no streams, use no supplies of water. They simply sit in the desert or out at sea, soaking up the power beam. They involve no unsightly structures. If a rectenna were built off the New Jersey beaches, all a beachgoer would see would be a line of buoys or marking lights to warn ships to steer clear. If the rectenna were two or three miles offshore, he would not even see that.

Power satellites built in a space colony offer more than this. They may be the key to overcoming the catastrophe predicted in *The Limits to Growth.*

One of the major criticisms directed against the original Forrester model was that it treated the world as a single entity. In reality the world has a relatively small number of industrialized nations, in Europe, North America, Asia, and the Soviet bloc. There are a larger number of underdeveloped nations, in Latin America, Africa, and much of Asia. Dieter R. Tuerpe of Lawrence Livermore Laboratory has developed a "two-sector" world model. The two sectors correspond to the developed and underdeveloped nations, with each sector modeled using the equations of Forrester's models. This corresponds to a highly Malthusian set of assumptions. Results from the model at least represent a worst case, so if a change in the model serves to make things better, then in the real world the situation would probably improve even more. Of particular concern are the prospects for the underdeveloped nations.

Tuerpe's two-sector model shows results which differ little from those of Forrester. The population continues to grow until late in the next century. The available food per capita, not very high today, diminishes to below the level of the year 1900. The quality of life falls to abysmal levels as does the standard of living. By the year 2100, conditions throughout the underdeveloped nations are perhaps somewhat poorer than exist today in Calcutta.

Peter Vajk (rhymes with "like"), of Science Applications, Inc., has extended Tuerpe's model by incorporating a third sector. This sector represents *space colonies.* The colonies then are regarded as interacting with the rest of the world, producing power satellites and selling their electricity to the world's nations.

In Vajk's model major construction on the first space colony begins in 1982, using NASA's space shuttle launch vehicle, which will be available in 1980. The shuttle is far from being the least costly launcher which could be built for a space-colonization program. The assumption of its use leads to an estimate of $178 billion as the cost of the project. The schedule assumed for the project is not based upon providing the greatest amount of energy to Earth at the earliest possible date. It is based upon paying off the investment, or funds expended upon the project, from sales of electricity in the United States alone.

Vajk's schedule is as follows: The first colony is finished in 1988. Thereafter, using lunar resources, it can build another colony (a duplicate of itself) in two years or two power satellites in one year. Each colony holds 10,000 people, and each power satellite delivers 5 million kilowatts of power to the earth. By 1998 the number of colonies has increased to sixteen and they are turning out thirty-two powersats per year. By 2007 the investment is repaid and all revenues are assumed to be plowed back into building more colonies or more power satellites. It is assumed that the colonies are not allowed to go into debt to provide a more rapid economic return. Revenues are obtained only from the sale of electricity from the powersats with the price initially at 1.5 cents per kilowatt-hour, dropping with time to 1.0 cent. This is at least as cheap as the cheapest electricity available today and far cheaper than electricity now available in the developing nations. Nevertheless, Vajk's model provides a 40-year transition period, during which power satellites grow to become the predominant source of the world's energy.

Vajk did this work on his own, in the manner of bootleg research, without a contract or formal support, while working for Lawrence Livermore Laboratory. It was in this fashion that he programed his computer and ran his equations. The results were startling.

According to Vajk's model space colonization will almost completely solve the developing nations' problems of limits to growth. The population, instead of continually rising, will level off and then decline. The standard of living will stay at or above the level of the present. The quality of life will also stay close to its present value. The food available per capita will show a slight decline around the year 2000 then increase steadily as the century progresses.

The decrease in population growth in this work does not follow from a siphoning-off of population as humanity heads for the rocket transports. Instead, it results from the rising standard of living brought about through cheap energy from the powersats. In Vajk's projections, by the year 2020 the population is 3.55 billion in the underdeveloped world, 1.33 billion in the developed, and 0.03 billion in the colonies. This compares with results from Tuerpe's two-sector model, without space colonization: 4.39 billion in the underdeveloped world and 1.82 billion in the developed. In Vajk's model by 2020 the world's population is leveling off rapidly, with the worldwide increase being only 12 million people per year. This compares with the present-day increase of close to 100 million per year. Of the population increase in 2020, half is leaving Earth for the space colonies.

The nations which build the first space colonies would wield immense power in such a world. They could become the future equivalent of today's oil cartel if they wished. The space colonist, visiting Earth, might be regarded in the future as we today regard an Arab sheik. Underdeveloped nations would quite likely accuse the colonists of being colonialists. But these nations would be able to move rapidly to build rectennas. Dozens of rectennas could be built in the Sahara, in the plateaus and deserts of Asia, in clearings in the jungles of Africa or Latin America. Under the power beams, the earth could become a steadily more hopeful place.

Paradoxically, perhaps the greatest benefits of space colonization would go to those nations which are doing the least to promote space flight today. The situation then could be quite similar to that brought about, in recent years, with the advent of communications satellites. These have given the United States television programs from around the world and improvements in our overseas phone networks. But beyond bringing Muhammad Ali live from Manila, they have had little impact on U.S. communications. The reason is we already had phone and TV networks in place, using cables and telephone lines. Many of the world's nations lack such systems and they eagerly joined the international consortium, Intelsat. For them, it was the first chance to build any sort of modern system for TV and for international telephony.

If the United States were to undertake to build space colonies, it would gain far more than the power and prestige of being the world's energy supplier. It would also gain a major new industry and source of wealth. Mark Hopkins carried out several studies on this point while with the NASA study on space colonization in the summer of 1975.

Hopkins' work differs from Vajk's in that he was not trying to solve the problems of the world. Instead, he was concerned with providing a new energy source which would be used principally in the United States. He assumed that only one-third of the power produced would be sold abroad. He also tried to develop a program which would build the first space colony at minimum cost, so that the program would get more easily through Congress. He sought to provide electricity from the powersats at the lowest possible cost, a fraction of a cent per kilowatt-hour.

With the cheerful optimism that was a feature of the NASA study, he assumed that major studies would begin in 1976. (The fact that they will start at a later date does not change his results.) Like Peter Vajk he regarded 1982 as the year for major construction in space to begin.

His work then proceeded based upon a detailed program plan as developed in the course of the study. This plan calls for the use of rockets more advanced than the space shuttle, which can be built within the next few years. These rockets will serve to cut the project cost. His results show that until 1986 the major costs involve research and development, along with the costs of building a moon base and the construction shack. After 1986, costs are dominated by the expenses required for building power satellites. Hopkins found the extra costs needed for the first and later colonies to be rather small.

In the program the first lunar material starts coming up from the moon in 1987 launched from the lunar base. During the next two years the work crew enlarges the initial construction shack and builds the first power satellite. But this powersat is not used to provide energy to Earth. It is transported to the L_1 point (see chapter 8), 40,000 miles Earthward of the moon and directly above its visible face. This powersat serves to greatly increase the power available on the moon, so that the moon-miners can mine and launch material at a much faster rate.

In 1989 the work crew builds the first powersat for commercial use. In that same year, they begin constructing the colony. With this first powersat, the colonists begin the process of taking over the entire United States market for new power plants. By 1999 the takeover is complete: all new and replacement power plants in the United States are built as rectennas for space-generated power.

Colonists begin to arrive in 1994. They have all arrived and the first colony is completed in 1998. After that the colony works to build more powersats. Later they build additional construction facilities and by 2011 they have the second colony.

Hopkins found the total cost of the program, through completion of the first colony, was $106 billion. The net costs of powersats produced through 1999 is $26 billion. After that the powersats produce a very large stream of revenue—as much as $80 billion a year by 2008. The entire cost of the project, principal plus interest, is paid back by 2019. The greatest expense of the program in any year is under $8 billion, quite similar to the peak cost of the Apollo program when this cost is corrected for inflation. By this analysis the colonization of space resembles an Apollo program continued for twenty years.

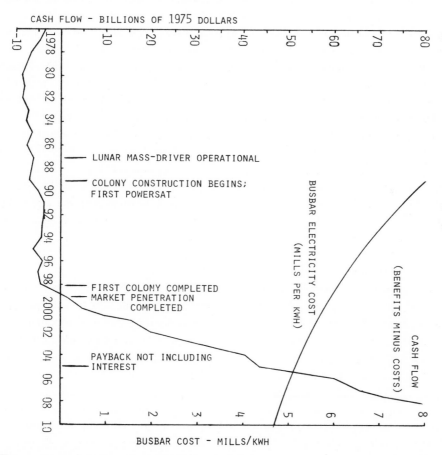

Costs and benefits of a program for construction of space colonies and of their use for building power satellites, following the economic analysis of Mark Hopkins. Dates indicate achievable milestones in the program, if it is pushed forward rapidly. Until 1999 there is a net cost of $5–$8 billion per year for construction and development. In the early years of the next century, revenues from satellite-generated power rise rapidly to $80 billion per year and more. Energy cost, initially 8 mills (0.8 cent) per kilowatt-hour, or half the cheapest price now available, drops in time to 3.5 mills. (Courtesy Mark Hopkins)

The cost of electricity from the powersats starts at 0.8 cent per kilowatt-hour, or one-half the lowest rate currently available, and drops with time to 0.35. This price is charged both within the United States and to customers abroad. Hopkins concluded that with foreign sales, the market could easily double in size, so revenues early in the next century could readily approach $200 billion per year.

What about the effects of inflation? Like most studies, Hopkins' work corrects for inflation by assuming that all costs are measured in 1975 dollars. If inflation makes the dollar of 2008 worth only half what it is today, the revenues then might be $400 billion, which is worth $200 billion in 1975 dollars. Similarly, the costs of electricity were given in 1975 cents. In addition, the economic analysis treated outlays for the program as debts which earn interest at a "real rate" of 10 percent. This rate is 10 percent plus whatever the inflation rate is in a given year. For instance, if inflation were 8 percent, the interest on the debt would be 18 percent, in the manner that interest rates are usually quoted. This very high rate of interest tends to make it quite hard for the project to turn a profit. The fact that in the analysis it nevertheless does so is testimony to the immense economic potential of space colonization.

At the 1976 Summer Study Dr. Gerald Driggers of the Southern Research Institute carried out his own analysis of the cost of space colonization. He was skeptical of Hopkins' and Vajk's results and expected that the colonization effort would be much more costly and difficult than had been envisioned; so he developed a construction schedule for the entire program. To his surprise, he found that the first powersat serving Earth could be built in 1992. His cost estimate for producing the first twenty 10-gigawatt powersats and building a lunar base, a space construction facility, and a colony for 6000 workers was $102.5 billion. At a subsequent press conference he stated: "I thought we would shoot down these earlier estimates, but they were right."

The foregoing studies offer a good deal of hope. There is the prospect of a major new industry in the United States, and of a solution to our energy needs. There is the possibility that the United States, through its space colonies, might become energy supplier to the world, overcoming the underdeveloped nations' limits to growth. From this, space colonization appears as one of the important new ideas to have come along in recent years.

It is possible that space colonization will become a generic subject, something like Ecology or Women's Lib, which everyone knows something about and has an opinion on. Then it will become part of political campaigns. Once an administration is elected with space colonization in the platform, or if Congress should pick up the idea, the twenty-year program begins.

Perhaps the president will appear before a joint session of Congress someday and present an address:

"I believe that this nation should set itself the goal, before this century is out, of establishing a colony in space which will be independent of Earth."

Chapter 5

FIRST OF THE GREAT SHIPS

The summer of 1969 was a halcyon time for NASA. It was not only the time of Apollo 11, which drew such enormous acclaim with its successful lunar landing. It was the time when NASA's administrators began to plan seriously for the post-Apollo future. In retrospect, it appears as a last summer of innocence, a last time before NASA would face the limitations of budget.

While launch crews prepared the next Apollo flights and astronauts intensified their training, a group of high-level Administration officials was charting the future of NASA's activities. This was the Space Task Group, chaired by Spiro Agnew. Agnew, who would later step down from the office of vice-president, came to Cape Canaveral on the morning of the Apollo 11 launch to announce that NASA's future goal would be to land men on Mars. Back in Washington, he led the Space Task Group in proposing a space program to exceed the Apollo program both in scope and in expense.

Its report, published in September 1969, listed three main program options for the 1970s and 1980s. All three called for the development of a small 12-man space station, a reusable space shuttle, a 100-man space base, and lunar orbiting stations as well as a station on the lunar surface. Two of the three called for the first manned expedition to Mars in the mid-1980s and projected funding levels of $8 to $10 billion per year for NASA by the late 1970s. The third option was less ambitious, giving no date for the first Mars landing but calling for simultaneous development of the space station and shuttle. This budget peaked at $6 billion for fiscal year 1977. By contrast, the peak NASA budget in the Apollo years had been $5.9 billion in 1966.

These optimistic projections seemed reasonable to a NASA elated with the success of Apollo. NASA officials believed they had an ally in President Nixon, who had often welcomed astronauts to the White House. This enthusiasm was not shared by the president's political and budgetary advisors, who argued that the nation's political climate and the competition for federal dollars would rule out such an ambitious space program. It was not long before NASA's hopes were put to the test. The agency had become accustomed during the Apollo years to receiving virtual carte blanche treatment of its budget requests. In the wake of the Space Task Group report, NASA was shocked when its fiscal 1971 budget proposal was rejected out of hand. The Office of Management and Budget slashed over $1 billion from it.

As the fall and the winter deepened and the decade of the 1960s came to an end, NASA officials began to lower their sights. In the Administration's judgment such projects as the space base, the lunar stations, and the manned Mars flight appeared too expensive to undertake in the foreseeable future. That left the space station and the shuttle.

Of these two, the space station was by far the more advanced in design. It was planned as the next step after the three-man Skylab program (Skylab flew in 1973–74), to fly in the latter part of the 1970s, carrying twelve astronauts who would conduct various programs of scientific study and Earth observations. In the course of the space-station design studies, NASA discovered that it needed a low-cost, reusable launch system. The launch vehicle to be used for Skylab would be the Saturn I-B, each flight of which cost $120 million. For Skylab only three flights would be needed, but the space station would require so many that much of the budget would be used up simply supplying the station. NASA proposed to develop the shuttle simultaneously with the station, at a cost of about $5 billion for each project. Agency officials described the shuttle and station as a single interdependent project. The space agency did suggest other uses for the shuttle: the launching and servicing of unmanned satellites, Earth observation studies, military reconnaissance, rescues in space. These were clearly secondary to its role in support of the space station and as a springboard to new manned adventures in space.

The "shuttle/station" concept was forthright enough, but it nearly killed both projects. Congressman Joseph Karth (D-Minn.) of the House committee on science and astronautics, usually an ardent supporter of the space program, claimed that NASA was seeking to extract from Congress a piecemeal commitment to what he called "its ultimate objective" of sending men to Mars. In the late spring of 1970, Karth introduced an amendment to block appropriations for the "shuttle/station." The amendment failed by the narrowest possible margin: a 53–53 tie. Early in July, Senator Walter Mondale introduced a similar amendment, which failed to pass by 32 to 28.

As a result, NASA quickly did an about-face on its justification for the shuttle. Instead of describing its use for new manned flights, it asserted that the shuttle could be justified in economic terms by the savings it would provide to an *unmanned* space program. NASA officials also "decoupled" the shuttle from the space station project, giving it a separate project staff and separate designation in the budget. The shuttle was now Shuttle, with a capital S. And they went shopping for other customers to use it.

They soon found that the Air Force could use it. The Air Force found itself in an unusual and quite fortunate situation: NASA needed Air Force business even more than the Air Force needed a new launch vehicle. As the secretary of the Air Force, Robert Seamans (later to head the Energy Research and Development Administration) told a 1971 Congressional hearing, "I cannot sit here today and say that the space transportation system is an essential military requirement." As NASA and the Air Force ultimately agreed, the Air Force would contribute its political support and its payloads, but would not put up money for Shuttle's development. The space agency, in its turn, agreed to design the shuttle to meet Air Force requirements. It would have a delta wing to permit increased maneuverability during re-entry and it would be designed to carry 65,000 pounds of payload to orbit in a cargo bay 60 feet long by 15 feet wide.

By late 1970, it was clear that this plan would succeed. Congress accepted NASA's new justifications for the program and the program was not again seriously challenged in the House of Representatives. In the Senate, Walter Mondale and a few others continued for some time to grumble that Shuttle was "a project in search of a mission." But on recorded votes, Mondale's motions to delete Shuttle funding went down to defeat in subsequent years by votes of 50 to 26, 61 to 20, and 64 to 22. In the spring of 1973, Senate opposition to the shuttle collapsed, Mondale declined to again offer his amendment to delete funding, and NASA's appropriation swept through to passage.

This did not mean, however, that NASA had a free hand in Shuttle's design and construction. Shuttle, born of compromise, would throughout its life be a creature of politics. The evolution of its design would ultimately depend as much upon budgetary considerations as upon technical requirements.

The shuttle has its roots in the dreams of the rocket pioneers of the 1930s. They foresaw the development of spaceships which would lift off from Earth, perform a mission in space, and return intact to Earth to be used for further missions. They scarcely imagined the actual situation of three or four decades later, when a rocket costing perhaps $100 million would lift off, perform a mission, and be destroyed either by falling into the ocean or by burning up re-entering the atmosphere. Arthur Clarke said of those days, "We were not that imaginative."

From 1940 to 1970, nearly all rocket development involved expendable boosters, either for use as ballistic missiles of war or for the first generation of space launch vehicles. Nor was there any sharp distinction between these uses. The United States found, following World War II, that after scraping the Nazi swastikas off the tailfins of the V-2, they could have a fine rocket for upper-air research. A generation later, NASA found that by scraping the Air Force insignia off the Titan III, it would have precisely the right launch vehicle to send its Viking spacecraft to Mars. But a few projects pointed to the future.

In the years after the war, the Air Force built the X series of rocket-powered research aircraft: the X-1, X-1A, X-2, and X-15. In the 1960s, the X-15 explored many of the aerodynamic and control problems to face a returning shuttle, carrying pilots to altitudes as great as 67 miles and at speeds above 4000 miles per hour. There was an active program of research in hypersonic wind tunnels which in the middle 1960s culminated in the programs of ASSET and PRIME. These involved aerodynamically controllable re-entry vehicles with stubby wings or with shapes capable of providing lift during atmosphere entry. Boosted by Atlas rockets, they were launched on suborbital flights to enter the atmosphere at 12,000 miles per hour and give data on flight at speeds and altitudes which the X-15 could not reach.

In January 1969, NASA awarded contracts for initial (Phase A) design studies of space shuttles. These studies were intended to demonstrate the feasibility of developing such vehicles; the contracts went to General Dynamics, Lockheed, North American Rockwell, and McDonnell Douglas. For all four companies the Phase A studies meant an opportunity to explore a world once reserved only to the science-fiction writers.

In June 1970, NASA awarded contracts for Phase B work, detailed design, to teams of companies headed by North American Rockwell and by McDonnell Douglas. This work was intended to produce specific, thorough designs of a two-stage fully reusable shuttle. But at the same time, the space agency took two other actions, which in time would lead to the shuttle as we know it today. It extended its Phase A feasibility study contract with Lockheed and gave new Phase A contracts to Grumman and to the Chrysler Corporation, directing these companies to investigate simpler, less costly design concepts. It also negotiated a contract with Mathematica, Inc., an economics research company in Princeton, New Jersey, to provide for a one-year study of the economic merits of the fully reusable space shuttle.

In late spring and early summer of 1971, the contractors presented their reports—and the roof fell in. It started on May 31, when Mathematica presented its economic study. On the surface, it seemed to justify the fully reusable design. Actually, it amounted to a strong warning against proceeding with it. Its calculations showed that any cost overruns would mean the design would fail to meet its economic goals. The question of overruns was tied closely to the risks involved in developing what in 1971 was a very advanced system. Both in NASA and in the aerospace industry, there were nagging doubts about the prospect of building the fully reusable design without encountering costly technical problems. The principal problem was the lack of flexibility in the design. If either the booster or the orbiter proved heavier than originally planned the extra weight could not be accommodated by a simple expedient such as increasing the length of a propellant tank. Instead, much of the system would require redesign.

NASA had also discovered in discussions with the Office of Management and Budget that the White House had no intention of allowing a space budget adequate for development

Two-stage fully reusable space shuttle proposed in 1969 by NASA's Max Faget. (Courtesy NASA)

Two-stage fully reusable space shuttle studied extensively by North American Rockwell and General Dynamics in 1970 and 1971. (Courtesy Rockwell International Corp.)

of the two-stage fully reusable design. The contractors had found that its development would require peak funding of over $2 billion. The Administration would permit a budget allowing only $1 billion for peak funding. With that discovery, it was back to the drawing boards for all NASA's contractors. As one aerospace executive put it, "Some people saw it coming and some didn't, but whatever the case we all knew by July that the whole damned system had suddenly gone up for grabs."

Seeking cost reductions, the contractors studied the possibility of developing the shuttle second stage (the orbiter) and flying it for a term of years atop some interim booster, a first stage already in use. They also tried the approach of using existing engines and avionics (flight instrumentation). These approaches offered some hope, but not enough. It was then in the fall of 1971 that the Boeing design teams had an attack of sheer inspiration.

These teams had designed the S-IC, the first stage of the Saturn V moon rocket. They had studied the possibility of using the S-IC as an interim booster. They proposed that the S-IC should be converted into the much-sought fully reusable booster instead of being an interim project. They proposed to attach large delta wings and a vertical fin to that first stage, to add a nose section with a pilot compartment and add landing gear and ten large jet engines. The jets would be placed together in a pod under the fuselage. They proposed to use the same engines being built for use in the B-1 bomber. When they completed their studies, they found they had a design with peak funding nearly a billion dollars lower than that of the two-stage fully reusable shuttle, yet with all the growth potential of the more advanced design and with far less risk in development. The peak funding was some $1.2 billion, only $200 million above the target. Once again President Nixon's budget officials said: Good, but not good enough.

Because it would fly back to Cape Canaveral after launch and because it used the same set of five F-1 rocket engines used in the S-IC, the booster was called the Flyback F-1. It attracted a great deal of support within NASA but the relentless requirements of the budget left it no hope. When NASA reluctantly abandoned the Flyback F-1, it also abandoned its last hope for a manned booster. As 1971 drew to a close, NASA and its contractors were studying unmanned boosters which would fall into the Atlantic and be fished out for reuse. There were both solid- and liquid-propellant designs to consider. Some years earlier, in the 1960s, NASA had built experimental solid rockets with diameters of 120, 156, and 260 inches. The 120-inchers found use with the Air Force's Titan III. The larger designs were kept around in case they might someday be needed. Various contractors had proposed to build simple, rugged, liquid-propelled "big dumb boosters," lacking guidance systems or complex pumps and plumbing. It was to this body of research that NASA now turned.

That was the situation in early 1972. On March 15 of that year the top administrators of NASA announced their decision on the choice of a booster for the shuttle in the U.S. Senate chamber. James Fletcher described the choice: the use of large solid motors, against the liquid-propelled big dumb booster. The problem, he said, was the booster would have to come down in the ocean and float until a tugboat could come to tow it back. The booster might sink. If it were the steel case of a solid motor, it would mean a loss of only $2 million, but if it were the liquid booster, it would be $75 million.

With that decision, the shuttle assumed its final form, except for rather minor changes (though they were not at all minor to the engineers who worked on them). The configuration of a delta-winged orbiter, with propellants in an external tank, lifting off with thrust from solid motors mounted on either side of the tank, has continued to this day. At that point, in 1972, it was only necessary to award the contracts in order to begin building the shuttle.

In a sense, the first major contract had already been awarded, a year earlier in July 1971. This contract had been won by Rocketdyne of Canoga Park, California. Rocketdyne was (and is) the nation's leading manufacturer of rocket engines, having built the engines for all three stages of the Saturn V as well as for most of the nation's other launch vehicles. The contract was for the space shuttle main engine (SSME), an advanced rocket motor burning hydrogen and oxygen at the unusually high pressure of 3000 pounds per square inch. High pressure was the key feature of the engine design since it would permit improved performance from a smaller motor.

In July 1972 the main shuttle contract went to North American Rockwell, now a part of Rockwell International, the automotive conglomerate. Perhaps the greatest irony is the world's first true spaceship will be built by what is now the Space Division of Rockwell International. This is no sleek, modern aerospace company full of Ph.D.'s and looking like a college campus. Its major plant facilities were built for aircraft production during World War II. Hundreds of engineers sit at desks in rows filling huge indoor work areas larger than a basketball court. There was a time when it was full of young scientists. In the years just after the war, North American was yeasty with new ideas—rocket propulsion, ramjets, inertial guidance, high-speed flight. Today many of these same scientists are at Stanford or MIT or the Jet Propulsion Laboratory, where they will cheerfully tell you that North American is a great place to be *from*. It is certainly remarkable that spaceships can now be designed and built as if they were new commercial airliners, that out of old airplane factories can come the stuff of long-held dreams. Chaucer would have understood:

> For out of olde fields, as men seeth,
> Cometh alle this newe graine from yere to yere.
> And out of olde bookes, in good feith
> Cometh alle this newe science that men lere.

In 1976, as the construction of the space shuttle approached completion, the question of a name for it arose. NASA officials preferred to call it *Constitution*. But Dick Hoagland, erstwhile science adviser to Walter Cronkite, had a different suggestion: *Enterprise*, after the "Star Trek" ship. He had a sufficiently close relation with one of the White House aides to have the idea put before President Ford. Further, through his contacts with the community of

"Star Trek" fans, 100,000 letters were sent to the president, urging that the name *Enterprise* be chosen. Ford acceded.

The *Enterprise* underwent final assembly in a plant at Palmdale in California's Mojave Desert. When complete, it was rolled out before the press and TV on September 17, 1976, to the accompaniment of a band playing the "Star Trek" theme. At the time of writing, plans called for it to be moved to Edwards Air Force Base for initial flight tests to begin early in 1977. In these, the orbiter will ride piggyback aboard a modified Boeing 747, being released to glide down to a landing at 200 knots on Edwards' 15,000-foot runway. Meanwhile, the second orbiter will be assembled and prepared for flight from Cape Canaveral. The first flight is scheduled for March 1, 1979.

The liftoff of a shuttle flight will be at least as spectacular as that of the Saturn V, which used to draw up to a million people to the Florida beaches to watch. There were only about a dozen flights, but in the decade of the 1980s there will be about five hundred shuttle flights. The shuttle will lift off with all five engines burning. The three liquid-fueled SSME's in the orbiter will burn with a pale yellow flame, as excess hydrogen in the exhaust flames in the air. The two large solid boosters will leave a smoky trail from the launch pad, extending far into the sky.

At 126 seconds into the mission, the spacecraft is at 142,000 feet. The solid boosters fall away, having propelled the shuttle to a speed of 4,715 feet per second. As the empty booster

The first space shuttle orbiter, Enterprise, *being rolled out at Palmdale, California, September 17, 1976. (Courtesy Rita Lauria)*

The complete space shuttle at liftoff. Propellant is carried in an external tank slung from its belly; two solid rocket motors are attached to this tank. All rocket motors fire at liftoff. (Courtesy Rockwell International Corp.)

casings fall, parachutes open, and the casings fall into the Atlantic 150 miles downrange, seven and a half minutes after liftoff. There they will float until collected by a tugboat, which will tow them back to Port Canaveral. The orbiter with its external tank continues its ascent. At 66 miles it enters a preliminary orbit at 25,786 feet per second. The external tank falls away to burn up in the atmosphere upon re-entry, and the orbiter is on its own. A short burst from its onboard orbital maneuvering engines places it in circular orbit at 110 miles. These engines also serve to maneuver the craft to any desired orbit, as high as 700 miles. There, the crew may launch or recover a satellite, using long manipulator arms carried in the payload bay.

The crew may stay up as long as thirty days. At mission's end, they use the orbital maneuvering engines to decrease the craft's velocity slightly. They enter the atmosphere with the nose held high, absorbing the heat of re-entry on the belly and the underside of the wings. These are covered with silica-fiber insulation. The leading edges of the nose and wing, exposed to temperatures as high as 3000°, are protected with carbon-fiber coverings. During re-entry, the crew may maneuver the craft as much as 1100 miles to the right or left of its orbital path. The landing is a critical time. Without power, the orbiter must land successfully on the first try and it is provided with a very advanced system for automated landing. It approaches the 15,000-foot runway at Cape Canaveral and glides in to land.

The orbiter is removed to the Vehicle Assembly Building, where the Saturn V rockets once were assembled. In two weeks it will be made ready for another flight. Like an airliner between flights, it will receive the attentions of a swarm of attendants, making minor repairs, overhauling or replacing its rocket engines, installing new payload, and replacing stores of food or oxygen used by the crew. Then it will be prepared for another flight. Cranes will lift it to a vertical position and a new external tank will be attached. Two solid motors, their casings filled with new propellant, will complete the assembly. The ship will move to the launch pad aboard one of NASA's huge crawler-transporters—the same ones used to carry Saturn V's to Pads 39 A and B. From one of these launch pads the shuttle will be fueled, counted down, and sent on its next flight. All will be routine, a matter of standard practice, and happening before 1980.

The shuttle will stand as the foundation for any space colonization effort. It will be used directly whenever it is necessary to return people or goods from the colony to Earth. Over a hundred passengers may squeeze into the payload bay for the return flight through the atmosphere. But most of the traffic will be outward bound, from the earth to the colony. Especially during the colony's construction, it will be necessary to lift 10,000 tons or more of payloads per year. The shuttle will lift over thirty tons on each flight at a cost of $10.5 million, or $160 per pound as the freight rate to orbit. While this is eminently suitable for a NASA program based upon the launching of more Landsats and Seasats, it is not adequate for colonization. The payload capacity is too low, the freight costs too high.

There is irony in this, recalling the situation late in the 1950s. At that time NASA was developing the rocket engine which would later be the F-1—a single motor with thrust of 1,500,000 pounds. Critics said, "You'll never use one of those, it's so large." And in a way they were right. When NASA began planning the manned lunar mission, it needed not one but five of these engines for the first stage of the Saturn V. Similarly, the shuttle as it exists today must be extended to carry five times its present payload.

The road to this result, however, turns out to be both short and simple. The shuttle, after all, is a collection of components: propellant tanks, engines, payload compartment, solid boosters and the like. These components will serve to create heavy lift launch vehicles (HLLV's). The process begins with a simple development. The SSME engines of the orbiter, instead of being mounted within the orbiter for return to Earth, must be packaged along with the avionics inside a heat shield so as to survive atmosphere entry on their own. This will provide a recoverable propulsion package which will re-enter the atmosphere following a launch and float to the ground with parachutes. Next the airplane-like orbiter must be replaced with a simple payload fairing to enclose the cargo on its trip up through the atmosphere. These simple engineering steps will produce an HLLV with two solid motors. Payload will increase to 150,000 pounds from the shuttle's 65,000. The freight rate will drop to $90 a pound.

This is not the end, however. The payload compartment, instead of riding piggyback on the propellant tank, can be mounted at the front of the tank. The tank then must be strengthened in structure to support the added loads. The number of solid motors is increased from two to four and the number of SSME's increases from three to four. These are encapsulated in their recoverable propulsion packages and mounted at the base of the propellant tank. Propellant feed lines are relocated. What results then is an HLLV with payload of 300,000 pounds and a freight rate of only $67 per pound. This is the launch system which the NASA study on space colonization, in the summer of 1975, recommended as the basic vehicle for use in space colonization.

Its use would put Cape Canaveral on an assembly-line basis. Two or three launches a week would be the standard. The four main assembly bays of the Vehicle Assembly Building would be in use round the clock as teams of rocketmen erected the propellant tanks, attached the solid motors, installed the cargo, and sent their handiwork to Pad 39 aboard the crawler-transporters. The tugboat operators out of Port Canaveral would expand their fleet to recover their hauls of spent booster casings at sea. Specialized Air Force flight crews would be on frequent call to perform aerial catches of the returning propulsion packages. All of this could be happening by 1982. A budget of half a billion dollars, less than 10 percent of the cost of the shuttle itself, will do to build the four-solid-motor HLLV.

The colonization effort will call for so much cargo to be lifted, so many HLLV flights, that it will be worthwhile to seek even further cost savings. Each use of a solid motor costs $2 million, or $8 million for the set of four. Much money can be saved replacing the solids with a fully reusable liquid booster. The Flyback F-1, rejected for budgetary reasons in 1971, will be quite appropriate. It will cost perhaps $5 billion and will take seven years to develop; it will not be available before 1985. But the combination of the Flyback F-1 and the main core of the HLLV, with its propellant tank and four SSME engines, will lift 400,000 pounds. The freight rate: a near-rock-bottom $25 per pound. The use of the Flyback F-1 will cut over $80 billion from what would otherwise be the cost of the colonization effort. Even this may not be the end. When it is again proposed to build a fully reusable liquid booster NASA will once again look at the designs submitted in 1971 by North American Rockwell and by McDonnell Douglas. One of these may give even greater economy than the Flyback F-1. The resulting launch vehicle could serve the colonization effort until well into the next century.

The payload compartment will be 27 feet in diameter and nearly 100 feet long. It will be suitable for a wide-body spaceliner to carry passengers and workers on their way to the colony. A partition down the center line of the compartment will produce two large passenger cabins, each with room for 100 or more people. Each passenger can be provided with a couch, on which to lie during the accelerations of launch, and a set of curtains to draw around the couch as in old Pullman railway sleepers. This will give everyone an enclosed cubicle in which to sleep during the three-day trip to the colony. The baggage allowances will be generous. A family of four will take at least a ton of personal belongings. There will be meals served aboard the spaceliner, which hopefully will be better than those served aboard airliners. There will also be rows of windows along the sides and the opportunity for passengers to experience weightlessness.

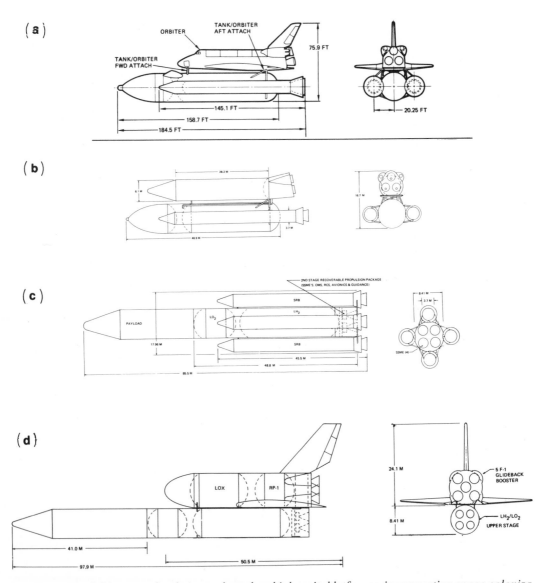

Development of the space shuttle into a launch vehicle suitable for use in supporting space colonization. (a) The baseline shuttle. (b) The airplanelike orbiter is removed and replaced by a payload fairing; the rocket motors are packaged in a recoverable re-entry body. (c) The payload is put at the front of the propellant tank; the number of main engines is increased from three to four; the number of solid rocket boosters (SRB's) increases from two to four. (d) The solid motors are replaced by the reusable Flyback F-1 first stage. The resulting launch vehicle can carry 400,000 pounds to orbit for a cost of as little as $10 million. This is the design selected as most promising for use in initial space colonization. (Courtesy Rockwell International Corp. and Boeing Aerospace Co.)

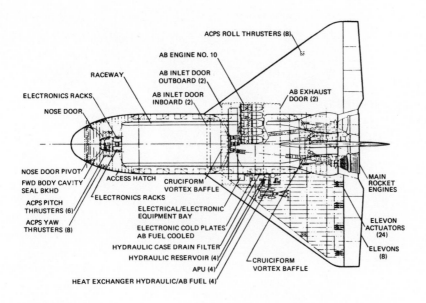

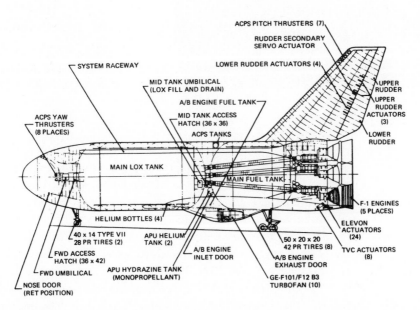

Details of the Flyback F-1 first stage. So far is this from being merely a theoretical or speculative concept that its designers are prepared to discuss what types of tires should be selected for the landing gear and where the auxiliary power units (APU's) should be located. (Courtesy Boeing Aerospace Co.)

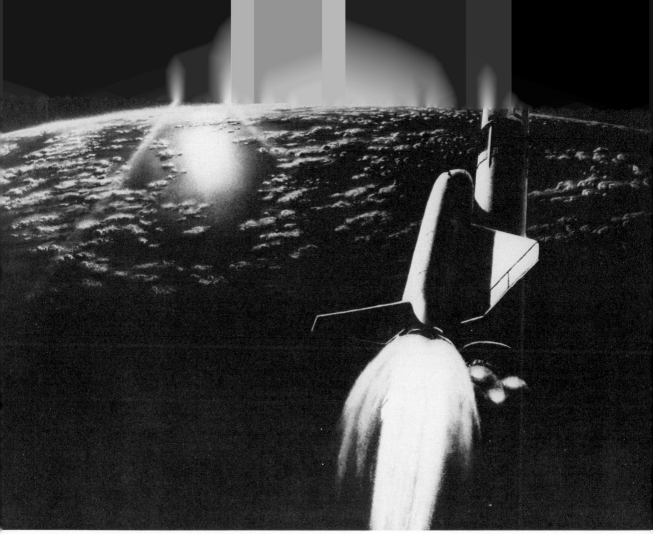

Ascent of a spaceliner derived from the Flyback F-1 and the launch vehicle configuration (d) *shown on page 72. (Courtesy Don Dixon)*

Much of the cargo carried by the HLLV's will be hydrogen and oxygen propellant in insulated tanks for use in transporting needed goods to the colony site or to the moon. The basic lunar transporter will be built around a single SSME which will serve to land 1000 tons on the lunar surface. Each flight will carry six standard payload containers, boosted to the moon with 4000 tons of propellant carried to orbit.

These projections illustrate the requirements for space colonization. Yet all these requirements can be met with launch vehicles derived from elements of the space shuttle or with vehicles such as the Flyback F-1 of 1971. It is for this reason that advocates of space colonization speak with such confidence. Any space project must necessarily rest upon the available rocket transport. This is as true for space colonization as it was, only twenty years ago, when scientists were planning the first Earth satellites. Rocket transport stands as a rock upon which space colonization may begin.

INITIAL ONE-WAY LUNAR TRANSPORT VEHICLE

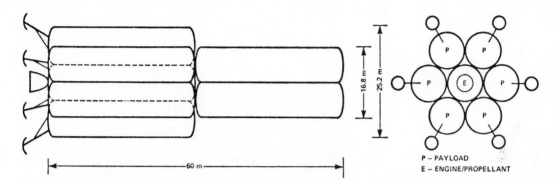

P – PAYLOAD
E – ENGINE/PROPELLANT

Lunar transport vehicle, whose propellant tanks are filled in Earth orbit with hydrogen and oxygen carried in launch vehicles similar to that shown in the preceding illustration. Its single space shuttle main engine allows it to land 1000 tons of payload on the lunar surface. (Courtesy NASA)

So the nation will proceed with the design, construction, and test flights of the shuttle. We will keep one eye on the budget and another eye on the need for a conservative justification. As for the shuttle, there is no doubt that we are truly building the first of the great ships.

Arthur Clarke said: "If man survives for as long as the least successful of the dinosaurs—those creatures whom we often deride as nature's failures—then we may be certain of this: For all but a vanishingly brief instant near the dawn of history, the word 'ship' will mean—'spaceship.' "

Chapter 6

THE MOON-MINERS

The moon was formed four and a half billion years ago in scenes of titanic violence. It grew from the assembly of many much smaller bodies. At first, these bodies—each the size of Mount Everest, say—assembled through collisions, at speeds no greater perhaps than a fast freight train. But in time these collisions built a core, large enough to have noticeable gravity, and then the impacts grew in violence.

The colliding bodies were not usually of solid rock; often they were mere loose assemblages, themselves formed from many smaller bodies. Many, perhaps, were of a porous or fluffy structure. But of whatever nature, when they struck the growing moon they exploded with the energy of a hydrogen bomb. The explosions dug huge craters in the surface, blasting material back into space. But some of the material clung to the moon, and in this fashion it grew.

The moon in those days knew only bombardments, explosions, and the impacts of flying rocks. Great craters and mountain ranges formed, only to be buried, a million years later, by the unending sprays of rocks flung from other, newer craters.

For a hundred million years this went on. Then, at last, the bombardment began to diminish. The reservoir of orbiting bodies, from which the planets were built, simply was running out. But before it ended, the moon was rocked by a series of cataclysms vaster than any it had earlier known.

The planetesimals did not all stay merely as mountain-size bodies. Some had themselves begun to grow through collisions, just as the moon's core had grown in its earliest days.

Eventually, some grew to diameters of a hundred miles or more. It was these bodies which now proceeded to strike the moon.

The first of them struck in the southern latitudes of the lunar far side. Its energy was that of a million of the largest bombs ever built, but did not come from explosive material. Rather, it derived from the sheer speed of so huge a body. The impact sent seismic waves through the entire moon. Huge masses of rock coursed outward, burying portions of the lunar surface thousands of miles away. In the near vicinity of the impact—for several hundreds of miles around—the moonquakes were powerful enough to fracture the rock of the moon's outer regions. Then titanic shock waves, radiating from the center, scooped up this thousand million million tons of rock and splashed it outward in an immense tidal wave, miles high.

When the shocks and flows of rock subsided, there was left a depression a thousand miles in diameter. Within it were three mountain ranges, concentric rings of rock, to mark where the tidal waves had at last spent their force. It was, perhaps, the greatest single feature ever formed upon the moon.

Yet it did not last. The final stages of lunar bombardment continued to build new craters, to erode existing mountains. When the bombardment finally ceased, the basin was heavily eroded, its rings of mountains worn almost beyond recognition; and so it remains to this day.

There were other great collisions, other surges of rock to dig immense pits upon the face of the moon. But at last the impacts ceased, for there were no more planetesimals. The moon rested.

Then slowly, over more hundreds of millions of years, heat built up inside the moon's outer crust, as radioactive elements decayed. While the earth's early oceans sparkled in the light of a younger sun, this heat melted rocks to lava. While blue-green algae formed matted accumulations in Earth's shallow waters (not for three billion years would their remote descendants advance the few yards to the beaches themselves), this lava accumulated beneath the surface of the moon. Finally, within the deepest basins and craters, it welled up. For half a billion years this welling continued.

Sometimes the vulcanism produced only small, local features, like the curious domes in the region of the Marius Hills. Other times it sent vast sheets of lava flowing slowly across the surface. Then again, a pool of lava would drain off through an underground tube. When the tube was empty, it would in time cave in, resulting in one of the moon's rills or canyons.

At last, the basins were filled in, and now even the vulcanism began to die. Yet there continued to be an infall of rocks from space. These were not planetesimals, now, but merely meteoroids, some larger than others. A few dug major craters, like Copernicus and Kepler. Most simply dug small pits and scattered small quantities of rock. In this fashion, the lunar soil became fractured and pulverized, to a depth of several feet.

For over three billion years there was no change. There was the endless cycle of searing, blazing sun and cold black night. More meteoroids fell, pulverizing the soil to deeper levels. Matter flowed out from the sun, and some of it implanted itself in the crushed stones of the

A planetesimal colliding with the moon. (Painting courtesy Don Dixon)

Formation of an immense lunar basin by impact of a body of some 100 miles diameter. The impact has produced a shock wave which is expanding across the lunar surface at several miles per second. Where it passes, it hurls mountain-sized chunks of rock outward, and leaves the surface it has passed glowing white-hot. The rock hurled outward will bury much of the lunar surface to a depth of many feet. (Donald E. Davis painting courtesy U.S. Geologic Survey)

lunar surface. Occasionally, a meteoroid larger than usual would fall, and would dig a crater. That was all.

These are the things the moon did not have. It had no water, not even water trapped in rocks. There was no air. Nor were there ores, or easily worked sources of metal. Only rocks, which might someday be made to yield up their content of aluminum and oxygen, iron and titanium—rocks, and the blazing, desiccating sun.

On the moon there was no life but on the earth, a quarter million miles away, an extensive community of life-forms was developing. In time, there arose upon the earth an animal with a rare sense of self-awareness. In time, some of its number came to look at the sky, and to learn to make tally-marks on bone. These saw the moon, and tallied now twenty-eight, now twenty-nine marks upon the bone. Slowly, dimly, the idea began to arise that there was order and regularity in nature.

Then one day, another object fell toward the lunar surface. Like many of the meteoroids which had fallen, it contained aluminum and oxygen. But unlike them, it was able to slow down so as to land gently on the surface. Within it were two living beings from Earth. They stepped forth onto the bright bleak surface and uncovered a plaque:

HERE MEN FROM PLANET EARTH
FIRST SET FOOT UPON THE MOON.
JULY, 1969 AD
WE CAME IN PEACE FOR ALL MANKIND

The first lunar visitors came as explorers and as scientists. They landed in their tiny cramped spacecraft and spent but a few hours before leaving. They set up instruments, made observations, collected samples. Some of them rode for several miles in lunar rover craft. When they departed, they left their footprints and the lower parts of their landing craft. If undisturbed by future visits, these will last for a hundred million years. Lunar conditions are that unchanging, lunar features are that permanent.

From what these first explorers have taught us, we now know much of the moon's early history. We know more: we know the nature and composition of the moon's rocks and soils. With this knowledge, we need not guess or wonder at what can be found there. We can use this knowledge as a foundation on which to build the colonies of space.

The fundamental idea of space colonization is the use of lunar and other extraterrestrial resources. With only the resources of Earth, there is no hope of building the colonies. With the resources of the moon, space colonies may soon become inevitable.

The lunar rocks are not rich ores, but they are adequate. They contain plagioclase and anorthosite, sources of aluminum. There is also ilmenite, for titanium and iron. Lunar rocks also contain silica—on Earth a useless waste product, but of value in space. They contain oxygen, too, chemically bound to the metals and the silicon. With the solar energy of space, it becomes possible to break down these rocks, to liberate their metals and their oxygen.

To build the colony, about 1 million tons of this rock must be shipped each year from the moon out into space. The task will be the project of a group of perhaps 100 people: the moon-miners. They will be the next group to go to the moon, now that the explorers have left it.

Their task will be straightforward: to gather the necessary lunar material and to launch it into space where it can be caught. To do the first, they will have bulldozers and power shovels built to operate in the lunar vacuum. To do the second, they will build a device called a mass-driver.

Their lives will be the harsh lives of workers at a remote outpost. They will be on the moon for limited periods—a year perhaps or eighteen months—and they will welcome the spacecraft which come to take them back. The work will often be monotonous and routine, especially after the initial construction is finished. For all that they will share the camaraderie that comes with being one of a few selected for important work. For it is on their work, more so than most, that the success of the colonization effort will rest.

One might ask whether it might be preferable to build a colony on the moon—a large comfortable place which would be more or less self-sufficient. Certainly something very much like this may develop in time. Over many decades large numbers of people may come to live on the moon as lunar operations expand.

But there are a number of reasons why that old science-fiction dream, cities on the moon, may be very slow in realization. For most purposes, free space is a better location than the moon for the building of a colony. To reach the moon, it is necessary to use rockets and supply their propellant. Not only is a colony in free space reached much more readily, but also it is much easier to bring large bulky items there. A colony can build huge, fragile power satellites and send them to another orbit; powersats could never be launched fully assembled from the lunar surface. The moon's gravity also interferes with the erection of immense but delicate solar mirrors or radiators, which are much easier to build in space. Finally, life on the moon will forever be influenced, and in many ways limited, by the two-week day-night cycle. This cycle may make it very difficult to grow crops. For all these reasons, the population in space colonies may be expected to grow much faster than the population on the moon. For a long time, then, the lunar activities will be carried out at a simple outpost or base, heavily dependent upon regular rocket flights from Earth.

Although space colonists will outnumber the moon-miners, there will be moon-miners before there will be space colonists. One of the early tasks in the space colonization effort will be the buildup of the lunar base. Only when it is functioning properly will it be possible to begin constructing the colony.

The first lunar flight in this effort will serve to found the base, to establish its location. In time a fair-sized city may grow up there, but to begin there will only be a collection of pay-load canisters from the rocket. This first flight will not land a fully assembled base, ready to support major operations. It will carry a small initial group of people together with what they will need to live on the moon for a while. Their task will be to survive, and begin the work which later flights will bring to completion.

This first flight will carry the initial items of earth-moving equipment. Among the most necessary of these will be a soil-blower, resembling a snow-blower, designed to scoop up loose soil and deposit it in a pile off to one side. There will be a tractor to pull heavy loads across the surface.

It is quite impossible to run this equipment with ordinary diesel engines. A diesel could be made to work in the lunar vacuum by supplying oxygen from a tank to the carburetor. But the exhaust from the diesel would prove difficult to restore chemically to its original state as diesel fuel. All of the lunar engines will burn hydrogen since hydrogen burns to produce water vapor, which is easily electrolyzed, recycling the exhaust to give back both hydrogen and oxygen.

For high efficiency the fuels should not be burned in anything resembling a diesel engine. Instead, they will be combined directly to produce electricity in fuel cells. These have been used since 1965 to produce power aboard manned spacecraft and they will serve as well at the lunar base. The first flight will also carry stores of hydrogen and oxygen for fuel.

One of the payload canisters will contain a most important item, to be handled with great care: the interim nuclear power plant. To bring a nuclear plant to the sunny moon may be seen by some as akin to bringing oil to the Arabs. There will be much interest in solar power for the lunar base but on the moon, the sun shines only two weeks of the month and nuclear power is available round the calendar.

The interim plant will be a simple affair of 100 tons mass. Even in the moon's low gravity, it will still weigh sixteen tons; so the tractor will be needed for its installation. With pulleys and winches, the work crew will lower it to the surface setting it on a wheeled chassis. Then the tractor will haul it a mile or so away. Other work crews will use the soil-blower to pile mounds of earth around its core for protection against radiation. They will set up a large radiator which gets rid of waste heat so that the reactor can generate electricity properly. They also will install electric power lines to bring the current to the base.

For living quarters, some of the payload canisters will be fitted out as a type of lunar quonset hut. The huts will be comfortable enough, the way a Holiday Inn is comfortable, but they will hardly cause the moon-miners to fondly remember their times there. These will also be detached from the landing craft and hauled to their locations. The work crews will then proceed to cover the huts with several feet of lunar soil, using the soil-blower. This will protect them against cosmic rays and solar flares.

The covering of soil will make it easier to keep the internal temperatures comfortable. In space temperature control is mostly a matter of exposing black- or white-painted surfaces to the sun, but on the moon it is made more complicated by the day-night cycle. A covering of soil will help protect the huts from extremes of temperatures. It is easier to heat a room than

The establishment of an early permanent lunar outpost. Part of a rocket propellant tank has been converted into living quarters and is being covered by lunar soil for radiation protection. (Courtesy Mc-Donnell Douglas Corporation)

to cool it and this is particularly true in space. So the moon-miners quite likely will set up sunscreens to keep their huts in permanent shadow. The soil covering then will act as insulation and heat will leak out only slowly. The heat from the miners' bodies and from equipment within the huts may act to maintain comfortable temperatures inside.

With electrolyzers to produce hydrogen and oxygen, powered by the nuclear plant, the beginnings of the lunar base will be complete. The population will eat the ordinary food of astronauts: freeze-dried meats and vegetables, Tang, Pillsbury food sticks, all reconstituted with water produced by the fuel cells.

The base will then be ready for expansion, prepared to grow into the major facility upon which so much of space colonization will rest. The growth will come with the arrival of other rocket craft, perhaps twenty in all, over a period of three or four years. Some of these will carry return rocket stages to rotate crews back to Earth. Most, however, will bring more moon-miners, more quonset huts, more equipment of all types, more food and fuel. They also will bring components of the two major lunar systems: the main nuclear plant and the mass-driver.

The main plant will generate 200,000 kilowatts of power, enough to supply a city of 100,000. Some of this will serve to run the base. There will be no lack of electrolyzers, gas liquefiers, communications gear and hot-water heaters. But most of the power will serve to run the mass-driver.

The mass-driver will be an electrically driven launcher, a sort of electromagnetic catapult. It will accelerate masses of lunar material to escape velocity, 1.5 miles per second. These masses are to be launched, one or two per second, day in and day out, indefinitely. They will fly out into space, curving in the moon's gravity and slowing down as they go outward. Two days after launch they will reach the catching point, 40,000 miles above the lunar far side. There they will be intercepted.

The lunar mining and transportation operation. At left, scoopers pick up loose soil from a pit, for delivery to the station at the right. There, the material is packaged into payloads and loaded aboard payload carriers ("buckets") to be accelerated for flight. The triple mass-driver provides this acceleration. (Courtesy NASA)

Nothing like the mass-driver has yet been built. Nevertheless, it rests upon such long-understood principles, such well-established engineering designs, that already we can describe it in great detail. These designs have been prepared for use in high-speed trains.

For a number of years, there has been considerable interest in building railroads which could compete for speed with airliners. In recent years, most attention has been directed to magnetic levitation and to propulsion by linear electric motors. It is this technology which will serve to build the mass-driver.

Such wheelless railroads (a term which may become as quaint as "horseless carriages") now appear to represent a transportation development as important as the auto or airplane. The first commercial trains of this type may be built in Japan. The Japanese National Railways were the first to demonstrate an experimental railroad of this new type, on June 25, 1972—a date which in the history of transport may one day rival the Wright Brothers' December 17, 1903, or Robert Goddard's March 16, 1926.

In such a train, as well as in the mass-driver, superconducting magnets lift or levitate the vehicle above a track of aluminum. The magnets produce no lift when the vehicle is at rest. However, at speeds of only twenty miles per hour the metal of the track no longer allows magnetic lines of force to penetrate freely. The lines of force tend to be rejected from the metal, thus generating lift. At sixty miles per hour this effect reaches nearly full strength; the

lines of force are almost entirely rejected, and lift approaches its maximum. At still higher speeds this magnetic repulsion changes little, but now there is a new effect. The repulsive levitation produces drag, but as the velocity increases the drag diminishes. This is quite in contrast to the situation pertaining to ordinary trains or airplanes, whose drag increases with increasing speed.

There are a number of ways to propel such a train. One of the most promising is the linear synchronous motor, which is to act on a powerful magnet aboard the train. A synchronous motor is a type of electric motor, and like all such motors it spins round, to produce torque. But such motors can be "unwrapped," or built in a straight-line design, along a track. Such a linear motor then produces, not torque, but acceleration along a straight line. A train, which is magnetically levitated and driven by linear synchronous motor, can be

The lunar base. A nuclear plant provides power for the mass-driver, which is fed with materials scooped up by vehicles such as the one in the foreground. (Painting by Pierre Mion, © National Geographic Society)

likened to a magnetic surfboard. The surfboard rides the forward slope of a traveling magnetic wave whose speed of travel and height are continually adjusted to keep the surfboard in smooth motion. The adjustments are carried out automatically, and give the description "synchronous" to the motor.

These developments will be applied to the mass-driver. At the cost of delivering some 20,000 tons to the moon for the lunar base, it will be possible to ship some 1,000 times this mass to the colony as unprocessed lunar material. The cost of the transport will be some fifty cents per pound and may be less. No method employing rockets can compete with this system, any more than air freight can compete with the cost of moving bulk goods by pipeline.

Both the mass-driver and the main nuclear plant will be shipped from Earth in modules of some 150 tons each. This is dictated by the lift capacities of the launch vehicles to be used. For the mass-driver most of the 150-ton packages will be loads of rail segments, or of sections for the trackside linear synchronous motor. The mass-driver will be built much as if it were a railway. There will not be spikes to pound or ties to lay, but there will be fabricated 100-foot lengths of aluminum track to lay, each weighing 400 pounds in the lunar gravity. These will be set in place upon the ground and welded together. Optical targets resembling bull's-eyes will be placed atop the track. When viewed with a telescope, sighting along the track, it will be possible to adjust the track alignment to ensure adequate straightness.

The main power plant will require much greater effort. Its sections, delivered in small groups aboard each new rocket to land, will be quite bulky as well as delicate. There will be generators, turbines, and systems for circulating water within the reactor. There will also be large radiator panels, monitoring instruments, safety systems, and the core of the main reactor itself. The total system will be some 10,000 tons in mass. Slowly, carefully, the assembly will proceed, with tractors and other equipment. These will be powered by fuel cells, the fuel for which will be produced with power from the interim plant.

In time, the last length of track will be laid, the last components of the power plant assembled. The major phases of lunar base construction will be at an end, and the moon-miners will be ready to supply materials for the colony.

Strictly speaking, it is not correct to call them miners, for they will not burrow beneath the surface. For the most part they will simply scoop up loose material lying at the surface, the fragmented debris of 3 billion years of meteoroid bombardment. They may have equipment for prospecting. This equipment, mounted on roving vehicles, will identify rock types by their response to electric or magnetic fields. But even the richest deposits will have only two or three times the metal content of ordinary soil.

A million tons a year will be the goal, the quantity of material to be gathered and launched. That is equivalent to digging out twenty football fields to a depth of ten feet. This is work for only a single small power shovel operating continuously. After twenty years, this activity still will not produce a crater visible in the earth's largest telescopes.

It is a curious thing, this view of humanity's works from a cosmic perspective. Men fancy that they have transformed the earth, and perhaps they have. Yet from only a few hundred miles up, most works of civilization are invisible, or are difficult to distinguish from geological features. So much more difficult will it be to make a noticeable impact upon the moon.

Not for lack of effort will this be true, however. Once in full operation, shifts of moon-miners will be at work at all hours. In the glare of the sun, they will unroll small aluminum shades and go about their work. In the lunar night, the soft bluish glow of earthlight will shine on them.

The soil and small rocks gathered will be brought to the central station of the mass-driver. There they will be bound together in forty-pound packages, each surrounded by a

Two-way traffic on the mass-driver. Buckets containing payloads are moving toward the horizon. The dotted line represents the trajectory of the payloads into space to the catching point, 40,000 miles above the lunar far side. Empty buckets are moving toward the foreground. They will be guided to the switchyard and the central station in preparation for another launch. (Drawing courtesy Don Dixon)

fiberglass bag made from lunar silica. There will be a small processing plant to make the crude fiberglass and the wrappings directly from lunar soil.

The central station also is an automated facility for servicing the vehicles which ride the track and which carry the packages. These vehicles are the "buckets." Each has a pair of superconducting magnets to support and guide the bucket along the track. It is these same magnets upon which the trackside linear synchronous motor sections will work to accelerate the bucket. There also is an array of mechanical restraints to hold the package during acceleration.

The superconducting magnets require the coldest substance known, liquid helium, which at −452° is only seven Fahrenheit degrees above absolute zero. When a bucket is in the central station, its main tank is filled with this frigid substance. During its last trip along the mass-driver, some of it will have boiled off. The gas passes around the outside of the magnets, to further insulate them, and runs into another tank aboard the bucket for storage. While still in the central station this tank is tapped and the helium drawn off to be reliquefied. A package of lunar material, in its fiberglass wrapping, is placed aboard and secured with four metal plates.

At this moment two or three dozen other buckets are in the central station, undergoing similar servicing. They all are guided on tracks, rolling on small wheels, so the central station actually functions as a railway switchyard. With precision timing each bucket is guided onto the main mass-driver track, like a car entering a freeway during rush hour. Once on this track the bucket quickly speeds up and levitates above the track. It is no longer supported by its wheels.

The initial acceleration lasts three seconds. In this time the bucket travels two miles. The acceleration is at 100 times Earth's gravity, and the speed is carefully measured with laser instruments. Then the bucket is at lunar escape velocity. Next, the bucket traverses an accurately aligned section of track wherein no acceleration is applied. The accurately aligned section permits the bucket oscillations caused by bumpy acceleration and a slightly uneven track to die out, the way a car shaking on a rough road becomes steadier once it hits even pavement. This is known as "passive magnetic damping."

The main restraining plates then drop away, and the package of lunar material is ready for release and for the start of its long flight. But it will not be released just yet.

Instead, the bucket and its payload fly out from the end of the track, in free space. They pass through a number of cycles of course adjustment, in each of which the bucket is tracked with laser beams so as to determine deviations from the planned trajectory. Then, magnetic coils apply trimming forces to the bucket, correcting for the errors. After ten to twenty such cycles of measurement, calculation, and trimming, the bucket and payload are within better than one part in a million of the planned velocity and position, and the payload is ready for launch.

At launch, while the bucket is in free flight, the last payload holds are retracted. The bucket, with the payload now floating inside, re-enters the track. The track then angles sharply downward, to snap the bucket away from the payload, a distance of six feet in 0.05 second (170 g's of acceleration). The reason for this is that the payloads themselves are slightly magnetic.

Lunar soil contains some magnetic iron, in a very fine-grained form which is difficult to separate out. Moreover, most common minerals exhibit a weak form of magnetism known as paramagnetism. Both types of magnetism mean the payload motion will not be completely independent of the powerful magnets on the bucket. These magnets may disturb the payloads' flight by only a few inches per second if the bucket is snapped away rapidly; yet even this error must be corrected.

So the payload passes through three more cycles of correction. The first of these involves a correcting coil at the start of flight, to take out most of the error introduced during snapout. The second and third cycles involve a slightly different method. At the correction stations, 2 and 100 miles downrange, electrons are sprayed on the payload, to charge it up to a good voltage—say, 100,000 volts. By laser tracking, required course adjustments are found. Then the payload passes between pairs of electrically charged plates, which adjust the motion in the same manner as electrons are controlled in a TV set. By the time all course corrections have been completed, the required launch direction, at a velocity of some 1.5 miles per second, is achieved to a precision of ten feet per hour in the velocity from side to side, or up and down.

The velocity itself, as well as its direction, may also be controlled to that accuracy, though it is more difficult to control the velocity accurately than to control its direction. But it turns out that this is not necessary because of a remarkable development in celestial mechanics.

By computing orbits to be followed by the payloads, it has been found that if they are launched from a suitable point on the lunar surface and aimed at a target, they will hit that target even if the launch velocity is slightly off. That is, the gravity of the earth and moon acts to focus the trajectories so they arrive at the target despite slight errors in velocity. The best such launch site is at 33.1° east longitude, near the craters Censorinus and Maskelyne. Then, the target can be chosen as a point in space 40,000 miles behind the moon, known as the L_2 point. A catcher or target located there will stay on station, since there the gravity of the earth and moon is cancelled by the centrifugal force due to the orbital motion of the catcher.

After snapout, with the payload on its way, the bucket enters a section of track where it is decelerated. Powerful sections of linear synchronous motor slow it down, at 300 times the acceleration of Earth's gravity, while recovering the energy of the bucket. When the bucket is below 100 miles per hour, it enters a banked and curved section of track. Now it will return along the backstretch of the mass-driver, to be guided to the switchyard and the central station in preparation for another launch.

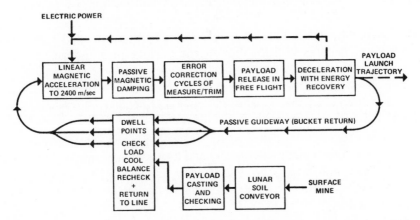

MASS-DRIVER (0.6-6.0 MILLION TONS/YEAR THROUGHPUT)

Schematic diagram of the main events in launch of a payload by the mass-driver. (Courtesy Gerard O'Neill)

NOMINAL PRELIMINARY MASS-DRIVER SECTION

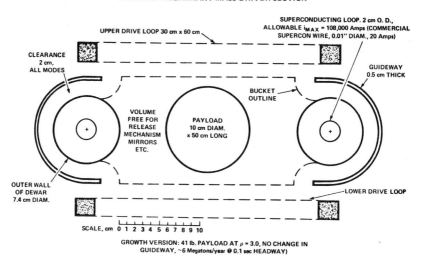

Cross-section of the track, accelerating coils, and bucket of the mass-driver. The track or guideway is C-shaped in cross section. The superconducting coil or loop on the bucket, cooled with liquid helium in the Dewar flask, acts both to center the bucket in the track and to interact with the drive loops (external to the bucket) which provide the force to accelerate the bucket. (Courtesy Gerard O'Neill)

Meanwhile, the payload flies onward, launched with astounding accuracy. If Garo Ypremian were to stand in the Oakland Raiders' stadium, near San Francisco, and kick a field goal between the uprights of Robert F. Kennedy Stadium in Washington, D.C., that would be the accuracy of launch. If Nolan Ryan, across the bay in Candlestick Park, were to pitch a strike to the plate at the Big A in Anaheim, 400 miles south, it would be the same accuracy.

The flight is initially nearly flat, a low line drive slowly rising above the lunar surface. Not for 200 miles (2 minutes of flight time) is the payload high enough to clear the lunar mountains. This is why the lunar base and mass-driver must be located on one of the broad lunar plains which were filled with lava 3 billion years ago. Thereafter, the rise into space is increasingly swift. Two days after launch, the payload is 40,000 miles behind the moon and approaching the catcher.

The catcher is a huge self-propelled craft, 300 feet wide and a quarter-mile long. During the years of the lunar base buildup this catcher and its mate (there are two) will be assembled in Earth orbit and moved to deep space. The catchers double as space ore carriers, transporting hundreds of thousands of tons of lunar material to the site of the colony. While one is in transit, another is on station.

Each catcher has a large rotating conical bag of Kevlar fabric. Kevlar is used in bulletproof vests—a nine-ply layer will stop a .44 magnum shell fired point-blank. This strength is needed, for the payloads come flying in at 600 miles per hour. They resemble the cannonballs of eighteenth-century naval warfare and catching them will prove a difficult problem.

There is a grid of cables across the front of the catcher. The package of lunar material strikes this grid and breaks up. This releases a shower of small rocks and gritty sand which flies inward. The material strikes the sloping sides of the Kevlar bag, bounces a bit, and comes to rest. The rotating bag holds it in place through centrifugal force.

From a distance the catcher resembles a sawed-off dirigible. The size and shape are similar and the resemblance is heightened by the seemingly tiny propulsion units on board. Yet each of these is a rotating tube, 100 feet long. Spinning rapidly, these "rotary pellet launchers" eject small pellets of rock, giving thrust to keep the catcher on station.

When the catcher is full it spins up its pellet launchers and slowly trundles off toward the colony site. It is now a true ore carrier, with loads and trip times quite like those of the supertankers of Earth's oceans. Like these supertankers, it is run by a small specialized crew who endure weeks of boredom during the main portions of the long hauls. These routine trips, more than any other flights in the space colonization program, will resemble the long, boring hauls of science fiction.

From the cargoes of these flights, from the megatons of rock and soil so carefully launched and transported, the colony will grow.

There will be moon-miners before there are space colonists. But once the colony is well

ROTARY PELLET LAUNCHER (RPL)

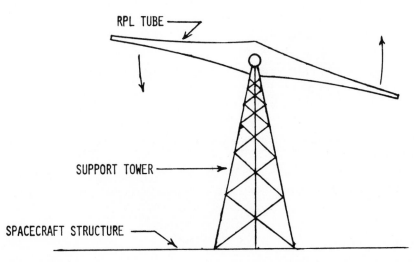

The rotary pellet launcher, which provides thrust by spinning rapidly so as to eject pellets at high speed. (Drawing by the author)

As the lunar base grows, it will become a center for scientific research. Richard Vondrak of the Stanford Research Institute has proposed that lunar craters be used to build radio telescopes such as the one at Arecibo. The group of three shown here can be employed separately for the study of different radio sources or can be combined to produce a single giant instrument. (Courtesy NASA)

The mass-catcher receiving a load of lunar material launched by the mass-driver. (Drawing courtesy Don Dixon)

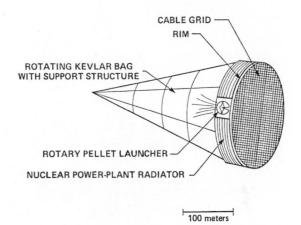

The mass-catcher. Payloads fly in from the right and break up on the grid of cables, thus releasing a shower of fine material which flies inward to be trapped by the rotating bag. Rotary pellet launchers, powered by an onboard nuclear plant, keep the catcher on station. (Courtesy NASA)

established, it will provide for an expansion of the lunar base by building a new power supply for it.

The colony will earn its economic keep by building power satellites. The first of these will be maneuvered toward the moon. Forty thousand miles above the near side, it will be stabilized in position, its power beam directed to the lunar base. There, solar power will at last replace nuclear power.

Nuclear plants will serve for the early years of the lunar base as a ready source of power. But it is no more desirable to rely on nuclear power for the long run upon the moon than upon the earth. Just as on the earth, solar power from a satellite will be the long-term supply which ensures the permanence of the lunar base.

Prior to the arrival of the power satellite, the moon-miners will prepare for it. They will build large trough-shaped reflectors of aluminum to concentrate and gather the microwaves to be beamed from space. Possibly they will build a small aluminum plant there to meet the needs of the lunar base and render themselves that much less dependent upon the earth.

They will not at first gather all the available microwave power since they will not need it. But as the years go by, they will build more and more trough reflectors. They will construct another mass-driver and yet another, to meet the growing needs of the space colony. As they expand, the power satellite will be there, providing the power they need.

Life at the lunar base will continue to be harsh. The moon-miners will continue to depend for many things, if not on the earth, on rocket flight from the space colonies. Not for a long time will they develop the comforts and conveniences of the colonies, and the moon-miners may continue to regard themselves as mere transient workers at an outpost.

For all this, during lunar night people will view the moon with telescopes and see something unusual there. In the extreme southern part of the Mare Tranquillitatis, there will be clusters of lights where before there have been no lights. There is a settlement on the moon.

93

Chapter 7

CONSTRUCTION SHACK

In astronautics there is an important concept which has stimulated and guided much work and research. It is an old concept, older than the first liquid-fuel rockets. It was worked out in fair detail as long ago as 1929 in a manner that even today would be regarded as essentially correct. Many of the leading pioneers in astronautics have proposed their versions of it, including Hermann Oberth, Wernher von Braun, and Krafft Ehricke. It has been studied in great detail under the well-known system of hiring a thousand ordinary engineers to do the work of a few men of genius. At least twice in NASA's history, it has come close to being the focus of the nation's space efforts. Yet it has never been built and never flown. The concept is the space station.

Space station, space base, manned orbiting laboratory—it has been studied under these names and many others, as well as under the three-letter acronyms (e.g., MOL, Manned Orbiting Lab; OLS, Orbiting Lunar Station) so loved by the aerospace industry. It is as old as the idea of space flight, as new as next year's NASA budget. It has been proposed to serve a variety of purposes but has never been built because there has never been a compelling need for it. Until now, perhaps.

The original discussion is in Hermann Oberth's 1923 work, *The Rocket into Interplanetary Space*. This small book set forth an agenda which spaceflight advocates will spend a century fulfilling. Oberth suggested that a station in space would be valuable for use in building large structures as well as for a refueling depot for spaceships. He wrote that stations could serve for observing the earth and aiding navigation: "The station would notice

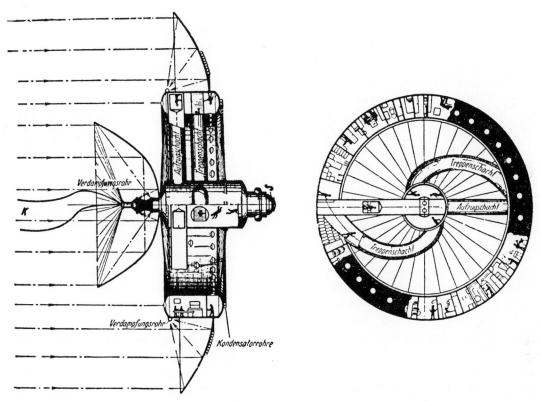

Space station designed in 1929 by Hermann Noordung. K *is the electric cable to an external observatory;* S *is the airlock;* Kondensatorrohre *are condenser pipes;* Verdampfungsrohr *are boiler pipes;* Treppenschacht *is a stairwell;* Aufzugschacht *is an elevator shaft. The design calls for power to be supplied by mirrors concentrating sunlight onto the boiler pipes. (Courtesy the Millikan Library, California Institute of Technology)*

every iceberg and warn ships . . . the disaster of the *Titanic* in 1912 would have been avoided by such means.''*

Six years later in 1929, an Austrian engineer writing under the pen name of Hermann Noordung gave a detailed design of a station. His ''Wohnrad'' (''living wheel'') was a wheel-shaped structure, 100 feet in diameter. It would spin to provide artificial gravity at 1 g, and it would obtain its power from the sun, focusing sunlight onto boiler tubes by means of mirrors. There would be a hub with an airlock, counter-rotating to stop spin and connected with large shafts to the rotating wheel. This was the first description of the type of space station made famous in the movie *2001*.

---

*This is not true, by the way. The *Titanic* had plenty of iceberg warnings but ignored them.

Twenty years later Wernher von Braun independently came up with a very similar design. Though he had never read of the Wohnrad, his station also was wheel-shaped, 250 feet across, and rotating to give ⅓ g artificial gravity. He proposed the wheel have two or three floors and the atmosphere be oxygen and helium at one-half sea level pressure. He also devoted attention to pumping water into ballast tanks so the station would continue to spin evenly as people went about their business inside, redistributing the masses within the wheel. Von Braun proposed to use this station as a base for the outfitting of expeditions to the moon and Mars.

His ideas received considerable attention. Walt Disney put together an hour-long TV program, *Man in Space*. *Time* put space flight on its cover (December 8, 1952) and the now-defunct *Collier's* ran a series of articles on von Braun's ideas, quite in the fashion that other magazines, a quarter-century later, would treat O'Neill's.

When the U.S. space program started a few years later, space stations ceased to be Walt Disney cartoons and began to be matters for public policy. Early in 1961 when President Kennedy was seeking a way to beat the Russians in the space race, he briefly considered the space station as the goal for the United States, before settling on the moon landing. Kennedy thought the Soviets might beat the United States in the race to build a station, but he felt we had at least an even chance of beating them to the moon.

In retrospect, it might have been better if Kennedy had picked the space station as a goal. All the major proponents of space flight had envisioned, after the initial orbital flights, the buildup of a space station as an essential prerequisite for any sustained program of activity in space. Instead, we bypassed this to go directly to building a manned lunar program. When this program could no longer attract public support late in the 1960s, the whole space program was left largely without a theme to keep it going. The result was NASA's rapid downslide.

Throughout the 1960s there were several major space-station projects, all of which in time were cancelled or cut back. Douglas Aircraft worked on its manned orbiting research laboratory (MORL) until the funding ran out. The Air Force had its manned orbiting laboratory (MOL), which actually got to the point of flying an experimental unmanned station to orbit before this program, too, was cancelled. Late in the decade North American Aviation and McDonnell Douglas churned out reams of paper space-station studies until Congress told NASA to have them quit.

For all this though, there actually was a project somewhat like a space station which was carried through and flew. This was Skylab. It never was intended as a space station in the classic sense as a permanent orbiting facility. But it was occupied, off and on, for nearly a year and periods as long as three months. It used ferry rockets to carry crews into space for their tours of duty, and they were able while there to do useful Earth observation and astronomical work. Had there been a space shuttle, and Skylab been in a slightly higher orbit not subject to decay, it is likely that it would have become a true space station.

NASA still is very interested in stations. One of the people in charge of that work is Jesco von Puttkamer, who is among the leaders of the bright-eyed visionaries in the space agency. He's originally from Germany, though he is too young to have been part of the Peenemunde group of World War II. He worked for von Braun in Huntsville, Alabama, then came to Washington to direct advanced planning for NASA's Office of Space Flight. He's the man who had primary responsibility for looking at Gerard O'Neill's work on space colonization.

Not only is he involved with space colonies, he also has been connected with "Star Trek." The Star Trek people put out a set of blueprints of the starship *Enterprise,* along with a space cadet manual. Their first printing ran to 450,000 copies. A Star Trek convention in Chicago drew 15,000 people and a second one in New York a few weeks later drew 30,000.

Von Puttkamer went to the one in Chicago, taking graphs and slides to show his planning through the year 2000. They asked him to show his presentation. Five thousand people attended and he repeated the show two or three times.

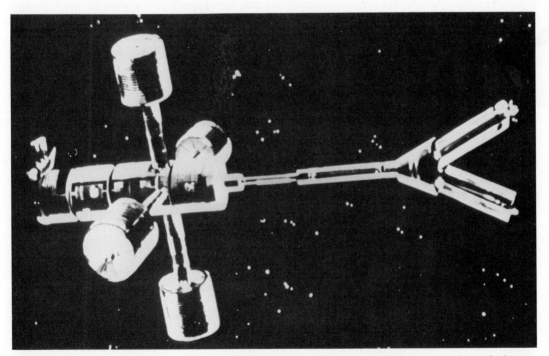

A space station design of current interest within NASA. The V-shaped structure at the right houses nuclear generating plants. The structure resembling two crossed barbells is the crew housing; it rotates to provide artificial gravity. Manufacturing operations are conducted in the central cylinder. (Courtesy NASA)

It is von Puttkamer who is responsible for NASA's current interest in space stations. He envisions a facility for 200 people in geosynchronous orbit. It would be used for assembling power satellites and for development of what he calls "space industrialization." He expects it could be ready by 1983 or 1984. In March 1976 his office let contracts to Grumman Aerospace Company and to McDonnell Douglas to study the concept. Such a station could be the next major space project after the shuttle, and von Puttkamer sees it as an essential first step toward a space colony.

Should it go forward, it almost certainly will undergo the same sort of changes and transformations that the space shuttle experienced on its way to becoming an official program. Von Puttkamer's recent award of study contracts appears to put the space station in about the same position that the shuttle was early in 1969. It will probably be two or three years before any firm designs are established. In the meantime, it is appropriate to describe some space-station concepts which are particularly applicable to building a space colony.

The particular space station to be used for this purpose is usually referred to as a "construction shack." "Construction" refers to its reason for existing and "shack" is a good description of the on-board amenities. It can house over two thousand workers and affirmative-action hiring procedures will ensure that about half of them are women. The station will be built in low Earth orbit then transported out to the site of the colony. Its construction and transport will take place during the same years that the lunar base and mass-driver are built up. When the first mass starts coming up from the moon, the construction shack will be there to work with it.

Much of the basic work in designing the construction shack has been done by Gerald W. Driggers of the Southern Research Institute, Birmingham, Alabama. His work draws heavily upon earlier space-station studies.

Driggers' construction shack will house 2232 workers. It is built around 36 modules which provide room for the crews to live, eat, sleep, and enjoy themselves as best they can. Each module will be 3 stories high and 50 feet in diameter, providing as much room as 2 or 3 four-bedroom suburban homes. Suburban homes might house at most a dozen or so people; each module on average will hold 62.

The modules will be arranged in groups of 3, each group providing for 186 people. The available space for these 186 is to be broken down like this:

Quarters, 10,100 square feet, or 54 per person
Wardroom and gym, 710 square feet
Galley, 316 square feet
Bathrooms and showers, 475 square feet
Medical area, 316 square feet
Laundry facilities, 118 square feet
Storage, 553 square feet
Dining area, 790 square feet

Although this is not exactly the same as gracious suburban living, it still is rather better than life on an aircraft carrier or submarine. There, crewmen sleep in bunks stacked one above the other with less than 10 square feet per man. Fifty-four square feet is about the size of a bathroom with toilet and sink only; if the men and women of the crew are accommodated two to a stateroom, the quarters will be about the size of dormitory rooms on college campuses.

Driggers proposes that there be two construction shacks, one to do the extraction of metals, the other to do the actual construction. Each will have eighteen modules, the modules being arranged in clusters of nine at the ends of large spokes. The spokes run through a central core which is equipped with airlocks and a docking port. This core, in turn, has on its top a very large sphere over three hundred feet in diameter. This is the construction sphere and each construction shack has one. It is in these spheres that the main work activities take place.

There is one other major item which is required: the power supply. There is good reason to use solar power since there is no day-night cycle as on the moon. The construction shacks will be equipped with their own power satellite, generating 300,000 kilowatts of power and transmitting it a mile or so to each of the two shacks. The power satellite will be no small project—its solar mirrors will be over a half-mile wide.

The power plant will not only provide for the construction shack, it will serve as a demonstration of satellite solar power. In developing the power satellite as a source of energy for Earth, there is need for some method of testing the whole system in a way that does not involve building an entire 10-mile-square powersat launched from Earth. The power needs of the construction shack can provide the opportunity for an overall test. Years before the first powersat is built for Earth all its necessary elements can be tested and proved out in the construction shack power plant.

The whole construction consists of two shacks, each having a core with clusters of modules set out on the ends of spokes. Atop each core is a construction sphere with a rectenna. Feeding microwave power to each rectenna is a single powersat with two transmitters a mile or so away. The two shacks together weigh 7000 tons, with another 3000 tons for the power plant.

All this is assembled in low Earth orbit, 300 miles up, then moved into deep space to the site of the colony. The supplies and equipment for the lunar base are shipped in 150-ton lots by ordinary rockets which burn hydrogen and oxygen, 6 lots to a shipment. The components of the construction shack equal the masses of naval destroyers and their transfer calls for slightly different methods.

The transfer begins with the transport into low orbit of huge stores of hydrogen. Eight thousand tons of liquid hydrogen must be lifted, and it takes about two years to do this. Ordinarily there would be little hope of keeping it liquid for that time, but the construction shack is large enough to be fitted with its own hydrogen liquefiers. Then, when hydrogen boils off

The construction shack. There are two such shacks, one for fabrication of the colony and one for metal production, both receiving power from a small power satellite. Each shack consists of a central sphere, 300 feet in diameter, wherein industrial operations take place. Below the sphere are clusters of housing modules, at the ends of arms; the assemblies of arms and modules rotate to provide artificial gravity. (Artwork courtesy Don Dixon)

within the tanks, it need not be lost. When the tanks are nearly full, the rocket engines arrive. These are of a new type and they heat the hydrogen directly to a high temperature, letting it blow out a nozzle to give thrust.

In the 1960s and early 1970s, there was a great deal of work on rockets of this type. This was project Nerva, which was named not for the second-century Roman emperor but as an acronym for Nuclear Engine for Rocket Vehicle Application. It involved a nuclear reactor to heat a block of carbon to over 4000°; the hydrogen was to flow through this block. For the construction shacks, there is no reason for the rocket to be nuclear since the power plant can supply energy enough. The rocket engine will actually be a sort of carbon arc, heated white-hot by an electric current. The carbon arc has been around since the days of Michael Faraday, a century and a half ago, and its use represents just another demonstration of how little in space colonization has to be newly invented.

These arcjets, as they are called, each give a thrust of 3000 pounds. One is on each major item: the two shacks and the power plant. This will give them a leisurely acceleration, and it is not hard to fly formation. The performance is about the same as using an auto engine to run a ship, but in space such slow accelerations can build up. It takes only a couple of months for that small fleet to reach the moon's distance from Earth. Later it is a simple matter to adjust position and wait for the first mass-catcher to come over, full of lunar material. At this point, operations at the site of the colony are ready to begin.

The operations involve the most difficult technical problems to be faced in the whole space colonization effort. We know about the rockets we will need, the power systems, even the mass-driver and the lunar base. All these represent problems for which major parts of the solutions have been studied or designed, or in some cases actually built. No one has ever tried to build an aluminum smelter to perform in space.

The smelting of metals, their extraction from lunar rocks and soil, represents a problem quite different and novel. Our terrestrial experience with aluminum production or steelmaking will be of only limited use. On Earth, metals usually are smelted from their oxides or from other simple compounds. Then the metals can be extracted through essentially a single chemical step. For iron the ore is heated together with carbon and limestone. For aluminum a current is passed through a solution of alumina in molten cryolite. In these industrial operations, air and water are available free or at low cost, and disposal of wastes is not usually a problem.

In the construction shack, everything will be different. The "ores" are complex chemical substances similar to ordinary rocks or clays. Water will be available only in limited quantities and all materials will have to be recycled. It is entirely hopeless to use the processes of Earth directly, and the processes which are used must operate in an environment where even gravity must be manufactured.

There is, however, plenty of solar energy, both for power and for heating of materials. Even more important is cost; the processes need not give their yields at prices competitive on Earth. On Earth a process for making aluminum might not be economic unless it could meet the world price, about $1 per pound. In space, a $20 billion construction shack which produces a million tons of aluminum would be practical. The production cost would be $10 per pound, which still is 4 times cheaper than the alternative of hauling aluminum up from Earth.

Moreover, the production of metals will also give a valuable by-product: oxygen. Oxygen, chemically bound, represents about 40 percent of typical lunar rocks. This is not only enough to provide for the breathing needs of people in space, but will also furnish rocket propellant. Once metal-producing operations are underway, it will become much easier to conduct rocket flights into lunar space. Rockets bound for the moon or the colony will need less hydrogen fuel than they would need if they were nuclear-powered. The combination of an ordinary hydrogen-oxygen rocket, together with oxygen being available at the construction shack, gives better performance than a nuclear or arcjet rocket would, using hydrogen alone.

How to produce the metals? Suppose, for instance, we want to get a ton of aluminum. The "recipe" would be:

"Take ten tons of anorthosite. Melt in a solar furnace at 3200°. Add water and quench the melt to give a glassy solid. Allow to settle in a centrifuge; pipe off the steam from the quenching to a radiator to condense it to water. Remove the glassy material from the centrifuge, grind fine, and mix with sulfuric acid. Pipe to another centrifuge to separate off the aluminum-bearing liquid which has resulted. Mix with sodium sulfate and heat to 400°; pipe to still another centrifuge to allow the resulting sodium aluminum sulfate to settle. Remove it after it has settled, and bake at 1470° to produce a mixture of alumina and sodium sulfate; wash out the latter with water. Mix the alumina with carbon, and react the mixture with chlorine. This gives aluminum chloride. Put the aluminum chloride through electrolysis. Result: One ton of molten aluminum."

The first steps of this recipe are called the melt-quench-leach process and have been tested by the Bureau of Mines. It has succeeded in recovering over 95 percent of the alumina present. The treatment with carbon and chlorine is carbochlorination. This process, along with the electrolysis, was developed and patented by Alcoa. All in all, it is quite a difficult and roundabout procedure; yet it appears to be among the simplest available. For instance, it is easier to electrolyze chlorides than oxides. This is why the alumina, or aluminum oxide, is carbochlorinated.

The whole process also has to provide for the recycling of chemicals. Sulfuric acid is re-formed following the removal of the sulfate. The electrolysis recovers chlorine. Finally, the carbochlorination produces carbon dioxide, which can be "shifted," as a chemist would say, to produce methane by combining it with steam over a catalyst. When the methane is heated to high temperature, it breaks down so that carbon is recovered as a form of soot.

The initial step, melting the rock in a furnace, also releases small quantities of elements blown out from the sun in its solar wind, and implanted in the grains of lunar soil. Only traces of these elements will be found; yet the construction shack will be processing such huge quantities of material that these traces can add up to significant amounts. In 1 million tons of lunar soil, there are about 40 tons of hydrogen. This is sufficient to make enough water to fill an Olympic-size swimming pool and will be much appreciated. There are 100 tons of nitrogen and 200 of carbon, which will help the agriculture. There are over 500 tons of sulfur, to help keep up supplies of sulfuric acid, and 2000 tons of sodium.

Titanium will be easier to produce. It comes from the lunar rock ilmenite, which also contains iron, and can be separated easily from other rocks since it is magnetic. The ilmenite is heated together with hydrogen, producing water, iron, and titania (titanium dioxide). Iron comes off easily as a by-product. The water is electrolyzed to recover the hydrogen. The titania goes through the same carbochlorination and electrolysis used for the alumina. This gives molten titanium.

Glass, real glass for windows, is another byproduct of the smelting. After acid leaching, the material left behind is nearly pure silica. This is remelted in a solar furnace then shaped into panes of high quality.

The smelters will get their ores from the moon but will rely on Earth for their chemicals. The trace elements from the moon will be useful mainly to replenish supplies lost in processing. The smelter will need several hundred tons of chemicals, including water, sodium sulfate, sulfuric acid, chlorine, and carbon. When operating at full capacity (150 tons per day of aluminum), the aluminum smelter will weigh 7600 tons. It will use 115,000 kilowatts of power (40,000 of which are to recover carbon via the shift process) and require 1 million square feet of radiator surface for cooling.

Not all of the smelter will come from Earth along with the construction shack. To a large degree, the smelter will have to be built up in space. Some of the growth will come by shipments from Earth, bringing chemicals and specialized items of equipment. But to a large extent, the smelters and construction areas will grow the same way as the colony: from lunar-derived metals. At the construction shack, the first construction will not be on parts of the colony or even on power satellites. It will be directed to expanding the metal-producing facilities of the shack itself.

This represents a new and challenging topic, the construction of large facilities in space. There is already considerable interest at NASA in this. The prevailing view is that it will involve many routine, repetitive operations which can be carried out by automated machines. The roles of human workers would be those of supervisors, expediters, and troubleshooters.

While the conditions of space make smelting difficult, the same conditions make construction and assembly easy. In terrestrial construction projects the actual assembly or joining is not difficult; it is a matter of installing fasteners or of putting in rivets. The difficult parts all involve gravity. There are heavy beams which must be hoisted, carefully balanced, and swung into place; for the unwary worker, there is a thirty-story drop down to the street. But in space, even the largest structural sections can be handled by small machines or by individual workers. The high-steel worker of weightless space does not have to cross girders or beams with a sure-footed stride. He simply launches himself in the direction he wants to go.

It will not be difficult to develop TV-controlled mobile assemblers. These will remove structural units from a pallet, place each one in correct position, and make a weld to fasten it in place. For the welding, there will be charges of thermite at appropriate places on the

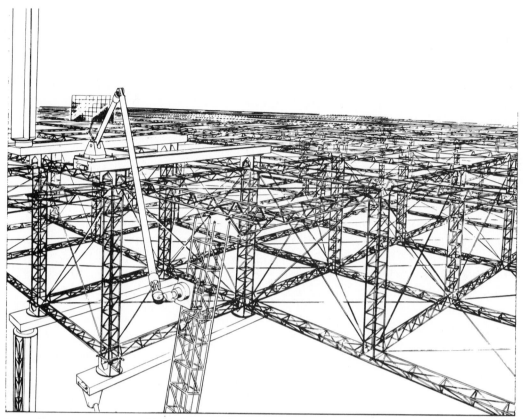

Use of an automated assembler in large-scale space construction. (Courtesy Martin Marietta Corp.)

structural units, fired by remote control. Thermite is useful since, like rocket fuel, it burns with an intensely hot flame, even in a vacuum. After completing one set of tasks, the assembler will move on command to another location to repeat the assembly sequence.

The assemblers will not be robots. They will be linked to small computers, programed to guide the assemblers through repetitive operations on command. For instance, there will be a computer routine which will direct the assembler to make a weld in a given spot; the operator only has to specify the spot. Rather than looking like robots, the assemblers will be similar to numerically controlled machine tools. Such tools have been used for many years. Some of the largest are at the Boeing Company to rivet and assemble jetliner wings without the need for human riveters.

The assemblers will work with fabricated subsections and parts. Some fabrication operations will change little from terrestrial machine shops; for example, the drawing of wire for cable. Hot and cold rolling of steel and aluminum plate will change little in space. Casting becomes easier in zero gravity.

There are some operations which do become more difficult in space. For instance, it is difficult to run a drop-forge. Machining of parts will produce metal chips and cuttings which will float around and wind up in odd places unless there are precautions. On the whole, however, zero gravity aids the fabricating of parts nearly as much as it aids their assembly.

In addition to these methods, there is an entirely new process which builds up large structures by squirting them out of a spray can. This is vacuum-vapor deposition. In this technique, a solar furnace heats aluminum not to its melting point but to the boiling point. The metal vapor squirts out an opening and is directed against a balloon-like form where it condenses. Provided the underlying surface for the aluminum spray is at room temperature, the condensed aluminum vapor will have the metallurgical properties of rolled and heat-treated sheet metal.

The simplest application is fabricating large seamless hulls, such as the main pressure shell for the space colony. First it is necessary to have an enormous balloon of the proper size and shape—a mile wide. This would be similar to, though much larger than, the Echo satellite of 1960 made of thin plastic film. The vapor-spraying facility would stand still while the balloon rotates past, receiving layer after layer of sprayed-on aluminum.

These are the kinds of processes which will serve to build the power satellites and the space colonies. They will need careful trial and development before they will be ready for use in space. The smelters can be set up on Earth and proved out before being disassembled for transport into space. The main problems there will involve corrosion, the operation of pumps and centrifuges, and avoiding leaks or major maintenance problems. All these can be checked out on the ground. Vacuum-vapor deposition can be checked out on space shuttle flights. Flights will carry a test balloon to orbit and a small vapor-deposition facility. The experience gained from this simple test will apply directly to the much larger task of space colony construction, just as when we paint a small wall we are confident we can paint a large one.

With the smelter operating, the construction spheres busy, and with its other facilities also in action, the crews of the construction shack will be prepared to start their work. They will have more to build than the colony, but the colony will be the focus of their greatest interest.

The construction shack will not be a very pleasant place to live. It will be more than just a work camp, and there will be opportunity for the crews to enjoy themselves when not at work. There will be microfiche libraries, videotape centers, and TV from Earth as well as frequent opportunities to make phone calls back home over the excellent communications which will be available. Nor will the work be particularly hard or difficult. Still, the people there will look ahead eagerly to the day they can move into the colony out of the cramped cubicles of the construction shack. That will mean the opportunity to eat lunch with a few close friends only, take a shower in privacy, or make love in a real double bed once again.

There will be little romance about the construction shack and even less adventure. Its inhabitants will be there simply to do the hard, necessary work of building. Nor will all of them wish to stay through the entire construction effort. Many of them will sign to serve a set tour

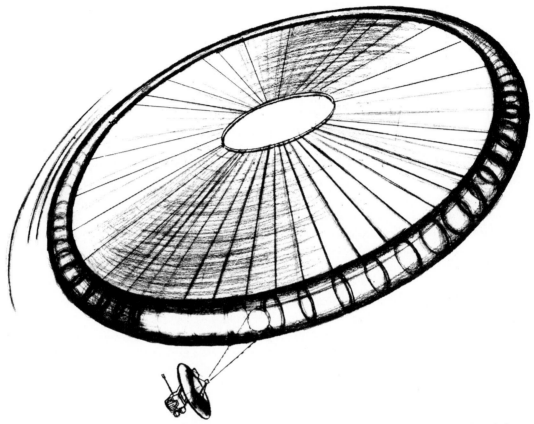

Construction of a space colony by vacuum-vapor deposition of aluminum over an inflated form. (Courtesy Don Dixon)

of duty and go back to Earth. There will be a regular system of shuttle rockets to rotate crew members home to Earth; and those who go will sing the song of Kipling:

> *We're goin' home, we're goin' home,*
> *Our ship is at the shore,*
> *And you must pack your haversack*
> *For we won't come back no more.*
> *Ho, don't you grieve for me,*
> *My lovely Mary Ann,*
> *For I'll marry you yit on a fourpenny bit*
> *As a time-expired man!*

So it will be for those who choose to return to Earth. But those who stay, those who commit their futures to space, will be the ones to inherit the colony which they build.

Chapter 8

THE HIGHEST HOME

Who will the space colonists be? What kind of work will they do and what skills will they need? To begin with, it is fairly certain that the principal types of jobs in the colony will be associated with large-scale construction or with running the colony's agricultural and life-support systems. To fill these positions the government will seek people with experience or backgrounds in shipbuilding, heavy electrical construction, industrial construction, chemistry, experimental agriculture, and plant science. There will, of course, be a number of professional and administrative positions in the colony. But these will be filled largely from the pre-existing project staff, by people who already will be engaged in the Earth-based design and development work.

It seems likely that preference will be given to single workers, male and female, under age thirty. Married couples, both of whom are acceptable for employment opportunities in the colony, will be welcome to apply. Facilities for child care will initially be limited, so preference will no doubt at first be given to couples with no more than one child. (As the colony becomes more solidly established, of course, many more children will be welcome.) Affirmative action hiring procedures will be in force.

This last feature will have special significance in regard to the ratio of men to women in the colony. It will be important that from the start the work force has approximately equal numbers of men and women. A work force mostly of men would soon give rise to a most unpleasant social atmosphere, resembling, if not a boom town—with its attendant crime, prostitution, and inflation—then possibly a prison or an army camp. A sexually mixed work force will ensure the development of a more normal society. The colonists will meet, fall in and out of love, marry or divorce or live together and establish homes and families.

Legislation can help to foster a stable society in the space colony. Congress should pass a law providing that those who build the colony can attain the common goal of American society—a home owned free and clear; that is, a home in the colony itself.

These policies will encourage the settlement of space not by opportunists or speculators, but by homeowning families.

Now that we have a fairly good idea of who the space colonists will be, what will their surroundings be like? Gerard O'Neill had some comments on this, when he testified in Senate hearings on January 19, 1976. At the time, Senator Barry Goldwater was hosting hearings on the subject of space industry before the Senate Subcommittee on Aerospace Technology and National Needs. When Gerry O'Neill came to speak, the hearing room was jammed. As part of his remarks, O'Neill said:

> It is natural for most people, and particularly for reporters and art directors, to become preoccupied with two features of orbital manufacturing. . . . One is, "Where is it [the space colony] going to go?" and the other is, "What is it going to look like?" I think the proper answer to the first question is "in an orbit high enough so that it almost never gets eclipsed,"* and to the second, "It will be a rotating pressure vessel, containing an atmosphere, with sunshine brought inside with mirrors." Beyond that, any further detail is almost certain to be wrong. For that reason, among others, I think it's unwise to get personally identified with particular designs. I'm for whatever works best, and it's too soon yet to be sure what that will be.

It is true that it's too early to say precisely where the colony will be or what it will look like. But it is not too early to say what would constitute a suitable location or shape.

For instance, a location on the moon is ruled out. Not only is there the problem of the two-week-long nights; the lunar gravity would make it very hard to build and launch power satellites. A low orbit around the earth like that of a space station could be tried. Such an orbit will stay in continual sunlight if it passes very nearly over the North and South poles and it is easy to reach it with rockets. But it is very difficult to transport lunar materials there, and it's a bit more difficult than it should be to send the power satellites out to geosynchronous orbit. If the power satellites are built from Earth-launched components, it will be well into the next century before this will be a paying proposition.

What about geosynchronous orbit? It is no harder to reach than any location near the moon, it is almost never eclipsed, and a colony located there would have no problems delivering the power satellites. What's more, it is already a much traveled orbit with satellites for communication and Earth observation. The founding of a colony there would be another instance in mankind's long history of establishing cities upon the main routes of commerce.

---

*"Eclipsed" here means "in Earth's shadow," i.e., shaded from the sun.

But geosynch orbit has two disadvantages, one minor, one major. The minor one is that it lies directly in the middle of one of the Van Allen radiation belts. While it is not too hard to protect colonies and rocket craft from radiation, it's better to avoid the Van Allen zones. The major problem is transportation of lunar materials there. Once launched to deep space by the mass-driver, the millions of tons of lunar ore cannot reach geosynch orbit easily. It takes slightly more than a mile per second of velocity change to get there. This is not hard for a rocket but to haul enough material to build the Great Pyramid of Cheops, it's too much.

Beyond these obvious choices, things begin to be difficult. The reason is Earth's gravity and lunar and solar pulls. These give disturbances to an orbit or, as an astronomer would call them, perturbations. The perturbations can be very important. For instance, some years ago there were some NASA officials who were interested in a space station to go into orbit around the moon. This was the LOSS, or Lunar Orbit Space Station. The interest declined abruptly when studies showed that after a few months it would crash into the moon.

In commenting on this, Robert Farquhar of NASA remarked in a report that the LOSS "would be a real loss." He then went on to describe an alternative which he called HOSS, Halo Orbit Space Station. This idea was based on perturbations helping rather than hurting in some orbits.

One of these orbits would take the colony two weeks to go around the earth—half the period of the moon. If you consider only the perturbations of the moon, the orbit will continually shift away from the moon. This is also true when the effects of the sun's perturbations are considered. Such an orbit is almost never eclipsed, and so is quite advantageous.

Prior to 1976, the most helpful perturbations, in the best-studied orbits, were thought to involve what are called libration points or Lagrangian points. In astronomy "libration" refers to a type of back-and-forth motion, like that of a pendulum, which bodies can undergo in the vicinity of these points. Libration points are locations where a spacecraft can be placed so it will always remain in the same position with respect to the earth and the moon.

Suppose the earth and moon were fixed in space and did not move. There would be a single libration point at the place where the gravity fields of the earth and moon cancel out. A body placed there would feel equal and opposite attractions from the earth and moon and stay fixed. If the body were moved a small amount, it would feel a slightly greater attraction from either the earth or moon. It would fall down, moving rapidly away from the libration point. Therefore, that point would be unstable.

But the earth and moon are in motion. It is the moon which does most of the moving. This means that in addition to the gravity of the earth and moon, account must be taken of the centrifugal force acting on a body in orbit. Libration points are the points where these *three* effects cancel out: the two gravity fields and the centrifugal force.

The French mathematician Lagrange showed in 1772 that there are five such points. Three of them lie on a line connecting the earth and moon; these are L_1, L_2, and L_3. They are unstable; a body placed there and moved slightly will tend to move away, though it will not usually crash directly onto the earth or moon. The other two are L_4 and L_5. They lie at equal distances from Earth and the moon in the moon's orbit, forming equilateral triangles with Earth and the moon. In the earth-moon system these points are stable.

The situation, however, is more complex than this. The sun is in the picture and it disturbs the orbits of spacecraft and colonies. It turned out (from an extremely difficult calculation done only in 1968) that with the sun in the picture, a colony should be placed not directly at L_4 or L_5, but in an orbit around one of these points. The roughly bean-shaped or kidney-shaped orbit would keep the colony about 90,000 miles from its central libration point. It may seem curious to speak of an orbit around a point. Actually, the colony would be in orbit about the earth, but the simplest way to describe the orbit is from the point of view of an observer sitting always at the libration point.

It is also possible to have orbits around the other libration points. These orbits involve libration in two or three dimensions. Sitting on the far side of the moon and watching a spacecraft in orbit around L_2, an observer would see the spacecraft make the same type of motion as a skywriting airplane making the letter O.

Robert Farquhar calls such an orbit a halo orbit. It bears somewhat the same relation to the moon as does an angel's halo to its head. The orbit is unstable, but Farquhar has shown that the solar perturbations on it are usually small and, by adroitly firing small rockets at the proper times, it is possible to stay in orbit. It was the basis for his proposal for the HOSS.

A halo orbit is not a very good location for a colony, because of the need to fire rockets regularly. But the L_1 and L_2 points are very good locales for two of the major systems which support the transport of lunar material. L_1 is a good spot for the lunar power satellite to beam down the power to the lunar base. The powersat will follow a small halo orbit, firing small jets occasionally to stay on location. Also, the region around L_2 is the right place for the mass-catcher. If the lunar base is on the great plains of the near side, then the material launched from the mass-driver can easily fly near L_2.

So it was proposed that the colonies be near L_4 or L_5. The mass-catcher was to be near L_2, and would shuttle between there and the colony. The lunar powersat would be at L_1. (To date, no use has been found for L_3.)

In the summer of 1976, though, several new studies were conducted which indicated that L_5 was not the best location for the colony. One question considered involved the transport of lunar material from L_2 to the colony. It turned out that with the colony near L_5, this transfer would need a velocity change of over 1400 feet per second. So there was interest in finding a colony site which could be reached more easily. This problem was studied by the simple method of mathematically letting a catcher depart L_2 and following it to see where it would go.

The resulting computer solution showed it would go quite close to a stable orbit around Earth with a period of two weeks. Further, it was found that if the colony were located in such an orbit, it could be reached from L_2 with velocity change as low as 30 feet per second.

EARTH-MOON LIBRATION POINTS

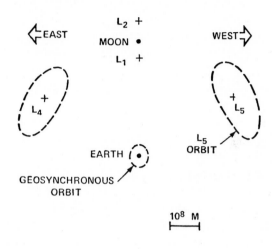

L_2 +

← EAST   MOON •   WEST →

L_1 +

L_4 +

L_5 +

L_5 ORBIT

EARTH

GEOSYNCHRONOUS ORBIT

10^8 M

L_3 +

TRANSFER TRAJECTORY
2:1 RESONANT ORBIT

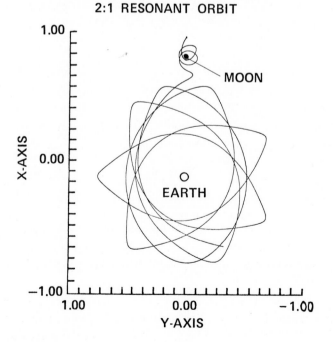

MOON

EARTH

X-AXIS

Y-AXIS

1.00 0.00 −1.00

1.00 0.00 −1.00

Trajectory followed by a catcher vehicle departing its station at L_2. It loops around the moon, then escapes into an Earth orbit with a period of about two weeks, an orbit known as a "2:1 resonant orbit." Such an orbit is the preferred one for the colony, since it can be reached from L_2 with very low velocity and also reduces the velocity required to reach the colony from Earth. (Courtesy David Kaplan)

When it was also found to be easier to reach such an orbit from Earth than to reach L_5, and also easier to deliver powersats from the orbit to geosynch, that clinched it. The colony will be in orbit some 200,000 miles from Earth at its farthest point and 100,000 miles at its closest point, taking a bit less than two weeks for each revolution.

Once a locale is chosen for the colony, the next thing is to consider how it may look. On Earth a home or apartment can have almost any shape desired, but in space only a few standard shapes are possible. These shapes, in turn, are dictated by the needs which the colony must meet.

The first of these is gravity. The colony must rotate to provide artificial gravity, in the sense of a "down" toward which things fall. This is not real gravity, of course, but is centrifugal force. Like many other artificial things, artificial gravity will not be a precisely identical substitute for the real thing. It will be good enough, but its artificiality involves some side effects which can cause problems.

The colonists will be Earth people whose bodies are accustomed to the sensation of gravity. There is still not a complete understanding of what happens in its absence, but what is known about weightlessness* is not completely reassuring.

Most of what we know about weightlessness comes from the experiences of the Apollo and Skylab astronauts. Shortly after reaching orbit, they often suffered from motion sickness. In Skylab, the motion sickness was especially bad, perhaps because of the increased room to move around in the rather large spacecraft. Fortunately, with the help of medication, the astronauts got over their spacesickness in a few days. After that, they had a greater tolerance than usual for motions that would ordinarily have made them want to reach for the little white bag.

The Skylab astronauts were able to live comfortably in zero gravity for the rest of their flights. However, after they got over their motion sickness, their red blood cell counts began to drop. Apparently there was a falloff in the production of red blood cells. Fortunately, after about a month in space, this loss of red cells leveled off, thereafter beginning to rise again on the longer flights. There was no increase in the body's destruction of these vital cells, but only an interruption in their production. By the end of the 84-day third mission, the astronauts' red-cell counts were nearly back to normal. Nor were changes observed in the immune response—the body's ability to fight disease.

There was one medical problem which did not level off and improve even on the longest Skylab flight—the loss of calcium from bones. The bones of the human body react to weightlessness the same way they react to prolonged rest in bed: by losing calcium at about .5 percent per month. This loss was continual throughout even the longest Skylab mission. The Skylab astronauts also lost nitrogen and phosphorus.

---

*A word is in order here about the meaning of "weightlessness" or "absence of gravity." It does not mean a part of the Universe where there literally are no gravity fields. The law of gravity is very hard to evade. "Weightlessness" means that an object is moving freely in response to whatever gravity fields are present. If it ceases to move freely, whether by rocket thrust or by standing on solid ground, there is weight.

The calcium loss is serious because it can weaken bones and make them more prone to fracture. This problem still has not been solved. Possibly it can be controlled by adding calcium to the diet and by having people in space perform exercises which not only give the muscles a workout but which also put stress on the bones.

The loss of calcium will not be a problem for periods of a year or so in space. It will be quite possible for crews to spend tours of duty in a weightless construction shack and very likely they can exercise or receive medical treatment to stay weightless for much longer times. But this can only be a temporary expedient. The colonists will certainly need to have weight.

It is desirable to make the colony as comfortable and Earthlike as possible. Space colony designers have tended to agree that the colony should provide normal Earth gravity, one g. We could consider lower levels, but we have no experience with living in levels of gravity between zero g and one g. The experience of zero g involves rather complex medical effects and clearly it is best to be cautious, to make no unneeded changes in the lives of the colonists.

The gravity will not be precisely the same everywhere in the colony. It will diminish toward the center. In a two-story home in the colony, gravity will be about .5 percent weaker on the second story than on the first. A 200-pound man would lose a pound just by climbing up the stairs. So it probably will be a popular thing in such houses to put the bathroom scale on the second floor. Even more weight loss will be possible by terracing the apartments with the topmost terraces 200 feet above the bottom. Then the variation in weight will be as much as 6 percent. For a number of reasons, the top terraces will be the most desirable real estate; but they likely will be most desirable to people on a diet.

However, there is a side effect to the artificial gravity which must be carefully controlled. This is the Coriolis acceleration, which is a property of all rotating environments. It is a tendency to change speed as you go inward or outward from the axis of a rotating object. On Earth it is the Coriolis acceleration which makes hurricanes swirl to the east in the northern hemisphere, to the west in the southern. It tends to make water draining from a bathtub swirl to the right in the northern hemisphere, to the left in the southern; but the bathtub must be kept very still for this to happen.

In a rotating colony, the Coriolis acceleration would have other noticeable effects. In a baseball game a batter will sometimes hit a long line drive which curves foul. This is because of the wind, but in a colony the same thing could often happen because of the Coriolis acceleration. Its tendency to change the direction of moving objects could produce other curious results. A colonist might be watering his lawn from a hose and see the stream curve over onto his neighbor's barbecue. Or he might be taking a shower and the spray from the showerhead would curve onto the floor. If the rotation is fast enough and the rotating area small enough, even worse things might happen. A man might go to the bathroom to answer nature's call and miss the toilet.

These rather unusual events would not really happen in any colony of reasonable size and rotation rate. But it is the Coriolis acceleration which gives rise to dizziness and motion sickness when a person is rotating rapidly. To avoid this, it is necessary to limit the allowable rotation rate.

At first, Gerry O'Neill believed the limit would be three rpm. However, colonists will be going to work in zero g, commuting back and forth between free space and their rotating home. While it is quite possible to become accustomed to continuous living at three rpm, it is likely that people would suffer from motion sickness if they had to go back and forth twice a day between such conditions and weightlessness. The 1975 Summer Study finally decided that the allowable rotation rate would be one rpm.

When you know the allowed rotation rate and the required level of artificial gravity, it is possible to determine the required size for a colony. When O'Neill thought the colony should give 1 g at 3 rpm, it would have a diameter of about 600 feet. To house 10,000 colonists with a reasonable amount of room for everyone, the colony would have to be a mile long. This was O'Neill's cylindrical design for a colony, with mirrors to reflect in sunlight, the design he had discussed in his original article in *Physics Today*.

However, there is no requirement that a colony be built in the shape of a cylinder. The only requirement is that it be a simple shape which can hold an atmosphere. The colony in space is a sort of balloon or inner tube, with air inside under pressure, and to keep its weight down, its shape has to be similar to the shape of a balloon or tire, as we are familiar with them.

Before settling on a shape for a colony, there is one other criterion to be met. The selected shape should give colonists, living on the inside, a sense of spaciousness. They should be able to look around and see not only the buildings immediately near them but also long distances hundreds of feet away. They should be able to walk outside of a building and feel they are out of doors, not merely that they have gone from a small room into a larger one. It will be even better if the colony shape allows not only long lines of sight, but also keeps some parts of the interior out of sight. These unseen parts would be either around the bend or else at a fair distance. This is important because people do not enjoy living in an area which they can take in completely with one glance. They need to be able to get a change of scenery.

We know from experience on Earth that if people are in a large enclosure with a roof high overhead, they regard themselves if they were outdoors and not necessarily in a large room. Baseball fans in the Astrodome and in other domed stadiums are not aware of the roof overhead unless a batter hits a pop-up. In San Francisco and Atlanta are the Hyatt House Regency hotels; these feature lobbies with roofs 200 feet high. It is difficult to be aware of the roof since there are plants and people at eye level.

While with the 1975 Summer Study on space colonization, I had a different demonstration that a very high roof is not really a roof at all. At Ames Research Center where we worked, there are several enormous hangars built in the 1930s to house dirigibles. Their roofs

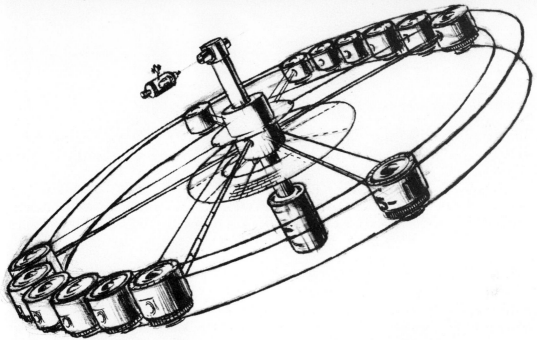

Modular space colony under construction. (Drawing courtesy Don Dixon)

are 100 feet or more above the floor. Within them are small offices, used by some of the people who work there. The offices are simple affairs of plywood nailed together. They have four walls, a desk, a telephone—and a plywood roof. The hangar roof does not seem like a real roof; for the people to feel that they really are inside an office, they have to have a roof directly over their heads.

The need for spaciousness rules out some designs that otherwise would be quite appealing. For instance, a colony could be built like a space station, from a lot of small elements arranged in a ring. Such a colony would be easy to build, but from the inside it would look like what it is: a collection of tin cans.

When our 1975 Summer Study got under way in June, we expected our colony design would look pretty much like O'Neill's cylinders. Thus it was a matter of considerable surprise when Wink Winkler, an undergraduate who was studying for admission to medical school, proposed that the proper rotation rate would be not three rpm but one rpm. This meant the whole colony design would have to change. When Winkler decided this in July, O'Neill was about to go to Washington to meet with Congressman Morris Udall and testify before the Congressional hearings. He did not exactly relish having to redesign his colony, especially on such short notice. So he sat down with Winkler and suggested a compromise: "Couldn't we do it at two rpm?" Winkler didn't say no and that was enough for O'Neill.

At first he thought of reviving an old idea, the "Bernal Sphere." In 1929 J. D. Bernal brought out an extraordinary book of scientific predictions, *The World, the Flesh, and the Devil.* In that book he gave the first description of the cosmic-ark method of interstellar flight. In this a self-contained spaceship would provide for generation after generation of people to grow and reproduce till the remote descendants would reach the destination star. The starship, he said, should be a sphere.

"Bernal sphere" design for a space colony. The sphere is the central structure; the structures resembling coils of hose are where agriculture is conducted. The disks at either end are radiators for waste heat. (Courtesy NASA)

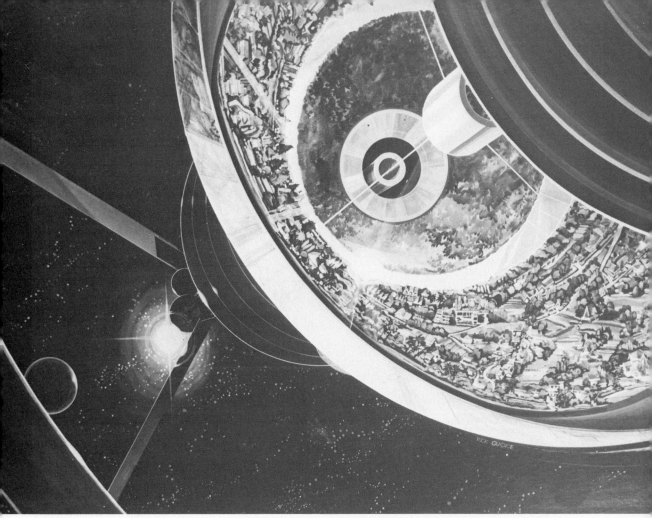

Cutaway view of "Bernal sphere" type of space colony. Some 10,000 people would live and work in the central sphere. A separate area, exposed to the intense sunlight of space, would be set aside for growing crops. (Courtesy NASA)

O'Neill adopted Bernal's ideas to furnish a design for a colony. But he preferred a cylinder to a sphere; so he settled at last on a rather stubby cylinder, 1500 feet across. He christened it "Sunflower": "Its petal-shaped mirrors, its tracking of the sun, its reliance on solar energy, and its property of being a warm habitat for life in the cold of space, all suggest the name." It was this design which O'Neill proceeded to describe at the Washington hearings:

> *It allows for natural sunshine, a hillside terraced environment, considerable bodies of water for swimming and boating, and an overall population density characteristic of some quite attractive modern communities in the U.S. and in southern France. It is startling to realize that even the first-model space community could have a population of 10,000 people, and that its circumference could be more than one mile. From the valley area, where streams could flow, a ten-minute walk could bring a resident up the hill to a region of*

Interior of "Bernal sphere" colony. The hang-glider pilot actually is engaged in powered flight, which is possible in the low gravity of the colony center. He pedals a bicyclelike arrangement which drives the large propeller at his back. (Courtesy NASA)

> *much-reduced gravity, where human-powered flight would be easy, sports and ballet could take on a new dimension, and weight would almost disappear. It seems almost a certainty that at such a level a person with a serious heart condition could live far longer than on Earth, and that low gravity could greatly ease many of the health problems of advancing age.*

Upon O'Neill's return, Wink was outraged that his recommendation of a one rpm limit should be treated so lightly. His careful research eventually convinced nearly everyone that the colony really should spin at one rpm.

A few evenings later about a dozen of the Summer Study people jammed into the apartment of Allan Russell, another Summer Study participant, to design a colony which would give one g at one rpm. Our architects were there as well as our structural designers and a number of others. The session lasted well into the evening. By the time we called it a night, we had the basic outlines of what would become known as the Stanford torus.

"Sunflower" space colony design. The pill-shaped central structure is where people live; it is topped by a solar furnace. Large mirrors reflect sunlight into its interior. Surrounding all is a ring for agriculture. (Courtesy NASA)

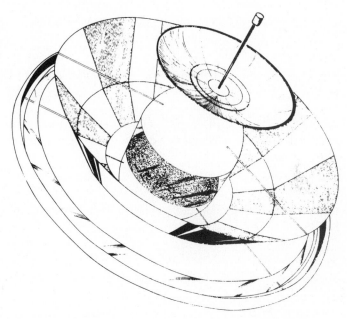

COLONY CONFIGURATION

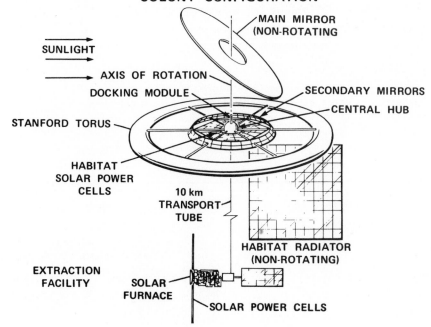

SUNLIGHT

MAIN MIRROR (NON-ROTATING

AXIS OF ROTATION

DOCKING MODULE

SECONDARY MIRRORS

CENTRAL HUB

STANFORD TORUS

HABITAT SOLAR POWER CELLS

10 km TRANSPORT TUBE

HABITAT RADIATOR (NON-ROTATING)

EXTRACTION FACILITY

SOLAR FURNACE

SOLAR POWER CELLS

"Stanford torus" design for a space colony. The colony proper is separated by up to ten kilometers from an extraction facility, but they are linked by a pneumatic tube. People live on the inside of the torus. (Courtesy NASA)

The spin rate and required gravity meant the colony would need a diameter of slightly over a mile. This immediately ruled out any sort of cylinder or sphere since it would then be too massive. Instead, we settled on a torus, an enormous inner-tube shape with the "tube" 400 feet wide and bent into a circle with a circumference of some 4 miles.

This design had many advantages. It was the lightest design yet studied. It offered the possibility of making effective use of interior space since there could be terraces for homes and agriculture. The colonists would have unobstructed lines of sight of half a mile and more; yet not all the colony would be open to view, since much of it would be concealed beyond the curve of the torus.

Once we settled on the Stanford torus, it was easy to design the rest of the colony. There would be a central hub. At its top would be a docking area, counter-rotated to stop spin, with ports for incoming spacecraft. The top of the central hub, which is 400 feet in diameter, would be the colony's "north pole."

The hub will be more than the physical center of the colony. It will have recreation areas and low-g gymnasiums and swimming pools. The maximum gravity there will be only about one-sixteenth g; so there can be human-powered flight. The hub will not only be a recreation area, it will be the main crossroads of the colony. All people and freight entering will pass through it. It will be the central area where six large spokes converge, each with elevator shafts, power cables, and heat pipes. Through this hub thousands of commuters each day will pass on their way to and from work.

Directly beneath the "south pole" of the hub is a de-spin system to which a second sphere is attached. This will be one of the construction spheres from the shack. It will be the

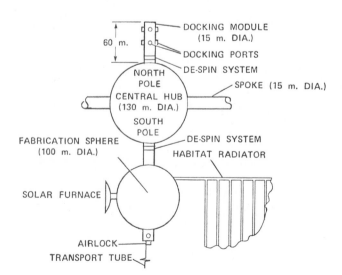

The hub of the "Stanford torus" design. It consists of two spheres, one to serve as the colony center, the other for fabrication. The latter has been transferred from the construction shack. (Courtesy NASA)

View of the "Stanford torus" space colony. The disc that hangs suspended above the wheel is a burnished floating mirror panel that reflects sunlight down onto slanted panels and into chevron shields that screen out cosmic rays. (Courtesy NASA)

fabrication area where major components are put together for the power satellites. To one side of the sphere will be a solar furnace. To the other side, edge-on to the sun, will be the colony's main radiator. This huge panel, a half-mile square, radiates the waste heat of the colony to space. Like the docking area at the north pole of the hub, the fabrication sphere and radiator do not rotate.

Below the fabrication sphere is an airlock and there is a long tubular passageway or transport tube. The tube runs to the second major industrial area, the extraction facility, where metals and other products are obtained from lunar material. Here is where the mass-catchers dock when they come from the L_2 libration point. The connecting tube resembles the pneumatic tubes used for mail service in some large buildings. A group of workers or a roll of sheet aluminum can go into a capsule; then air pressure sends the capsule along the tube in a minute or so. At the other end of the tube, air pressure brings the capsule to a gentle halt.

Directly above the central hub is a huge yet lightweight mirror over half a mile across. It floats above the colony with small rockets to keep its position properly adjusted. It is angled away from the sun to reflect sunlight down onto a double ring of secondary mirrors. These are mounted in rings around the hub, supported by the spokes, to reflect sunlight into the colony. The mirrors can be tipped or tilted to reflect sunlight away. This is done twice a day during the colony's dusk and dawn.

The spokes are fifty feet wide and go to the periphery of the colony. Not all the interior room is taken by the elevators, cables, and pipes which are so necessary to the colony. Together the spokes contain nearly as much room as in the Empire State Building. There are offices there and shops and laboratories for scientists. There, too, will be the University of Space.

The real life of the colony, however, is in the rim. There are terraced agricultural areas, homes, parks, and gardens. The inner one-third of the toroidal rim surface area consists of glass panels mounted on aluminum ribs to let in sunlight from the secondary mirrors. The remaining surface area is an aluminum hull of solid metal an inch thick, wrapped with cable.

Here is a city of 10,000, a small, bright island in the emptiness of space. Here will be the place where the human race first dimly glimpses its possibilities, its prospects, its future. It will be the true fulfillment of the dreams which once inspired a young and vigorous president in days when the world was newer and all things seemed possible:

"We choose to go to the Moon in this decade, and do the other things, not because they are easy but because they are hard. Because that goal will serve to organize and measure the best of our energies and skills. Because that challenge is one that we are willing to accept— one we are unwilling to postpone—and one which we intend to win."

Chapter 9

UP ON THE FARM

In space colonization, the really interesting questions do not involve things such as rockets or lunar bases or how fast the colony is to spin. They involve matters of how people will live in space. It is necessary to discuss the arrangements which will provide the colonists with a pleasant attractive life. O'Neill has written of lakes and gardens, of parks and of pleasant apartments. How do we know the colonies can actually be like this? How can we say that these prospects will ultimately be realized?

The implications of space colonization are so vast, its potential importance to the human race so great, that it deserves to be approached with care and attention to detail. It is necessary to discuss how the colonists get to the colony, how the colony can contribute to solving the world's problems, how long it would take to build one. It is equally necessary to look at such questions as breakfast, or what happens after the colonists put the garbage down the kitchen disposal.

How do you feed 10,000 people living in space? How do you provide the air they will breathe? These necessities must come from a space farm. Most people need about ten pounds of food, water, and oxygen per day. If the colony is to get its supplies by rocket from Earth, it will be necessary to launch one each day. Not only is this expensive, but it would contradict one of the major goals of space colonization: to build a self-sufficient human community.

There must be a space farm. And most if not all of the colonists' wastes will have to be recycled. The space colony will therefore have to be a closed-cycle ecology *par excellence*.

The farm must be small in area, since it will be a long time before colonies grow large enough to offer room for Kansas wheatfields. Yet it must provide an appetizing diet of variety and quality similar to that which we enjoy on Earth.

These goals will not be easy to meet. NASA and the Air Force have done research in space farming and closed ecologies for over twenty years. They have studied algae, yeast, and distilled urine as comestibles. Reports stating that the best strain of algae would be of the genus *Chlorella* have been issued. In 1960 it was counted a considerable step forward when the Air Force's Project Hermes succeeded in distilling from urine water which was alleged to be potable. (One Air Force scientist tried it and said, "It's no worse than some of the stuff you get at cocktail parties.") Mercifully, this line of research was abandoned and the Apollo astronauts got good fresh water made by combining hydrogen and oxygen in fuel cells.

For the colonists, there will be no algae except for feeding to fish and no yeast except when baking bread and cakes. They will have grain, vegetables, fruits, meat, fish, and poultry as well as eggs and plenty of milk or milk products. Steak may be a rarity and some of the people will miss familiar brand names. On the whole, though, there will be much less change in their diet than they would experience in moving to many places on Earth.

All this means high-intensity agriculture. In ordinary agriculture, Kansas style, a farmer may get 65 bushels of wheat to the acre, or about $1^3/_4$ tons. In Iowa, if the rain is good, you can get 140 bushels of corn to the acre, which is $3^1/_2$ tons. But since the middle 1960s, Richard Branfield and his associates at the International Rice Research Institute in the Philippines have consistently been getting 16 tons and more to the acre!

Branfield has been able to do this by taking advantage of several items which favor increased yields. The Philippines get much more rainfall than does Kansas and the growing season is longer. Branfield has been able to grow crops bred for shorter growing times. His rice, for example, matures in 90 to 110 days, instead of the usual 135 to 150. But his main techniques are interplanting and multiple cropping.

Interplanting means planting the seeds of your next crop before you have harvested the existing crop, planting the seeds between the already-growing plants. Crops grow slowly in the first weeks, while they are still seedlings. Interplanting overlaps the time of slow growth of your next crop with the time of rapid growth of the first crop just prior to harvest.

To do multiple cropping properly, Branfield has carefully worked out the proper sequences of crops to grow in succession. His best sequence starts with rice. Then he plants sweet potatoes and then soybeans; next corn and finally soybeans again. As a result, one of his acres will yield 2 tons of rice, 10 of sweet potatoes, 4 of soybeans, and 18,000 ears of corn. The stems and cuttings which are inedible by humans supply ten tons of forage for livestock. Such an acre will yield enough to feed thirty people continuously.

With these techniques, the whole colony of 10,000 population could live off the production of a single half-section, which is 320 acres. These methods are important much closer to home—they were not developed for space colonization. They were developed to feed people

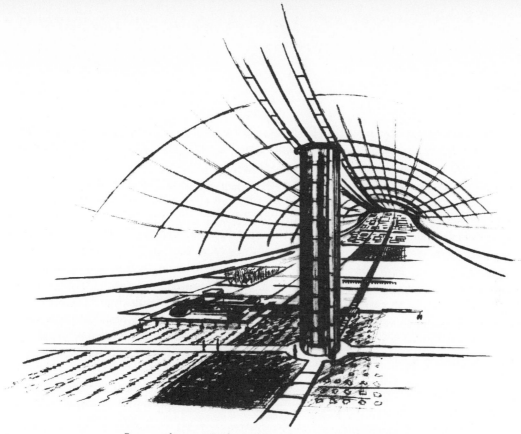

Space colony agriculture. (Drawing courtesy Don Dixon)

in Asia. Just as the technology of high-speed trains points the way to the lunar mass-driver, so do the methods of high-intensity agriculture show the direction to follow in planning the space farm.

Branfield's results, impressive as they are, can be improved on. It is possible to grow more. The Philippines have a nine-month growing season and less than ideal weather conditions. In the space colony, the growing season is continuous and can be adjusted to any conditions. We can control the temperature, the lighting, the moisture, the level of carbon dioxide in the atmosphere. Even if we give plants the best conditions they would find in the Philippines, but do so continually, the yield will be double what Branfield has obtained.

We should begin to regard the space farm as a continual producer of food. Rather than planting and harvesting at reasonably well-defined times in a well-defined growing season, we must imagine that at any time there will be some crops being planted, others newly growing, others ripening, and still others being harvested. Instead of thinking of yields in bushels or tons per acre over an entire growing season, we should think in terms of the yield as so many pounds per acre per day.

In Branfield's fields, the yields are about 125 pounds per acre per day. This includes not only food grown and eaten directly, but the daily production of meat from animals which eat the stems and cuttings as forage. By reproducing in the colony's farm the best Philippine days, the yield goes to about 245 pounds.

This is still only the beginning. In his book, *Photosynthesis, Photorespiration, and Plant Productivity,* Israel Zelitch of the Connecticut Agricultural Experiment Station reports yields exceeding 500 pounds for corn, sugar cane, sorghum, and millet, when these crops grow under optimum field conditions. In a laboratory experiment in England, plant growth initially at 340 pounds was raised above 1000 pounds by optimizing the atmospheric content of carbon dioxide at 0.13 percent. In the hydroponic garden at Arizona State University, John R. Meyers has grown forage under artificial 24-hour lighting, high ventilation, and controlled temperatures. His yield: an astounding 15,400 pounds per acre per day!

It is entirely reasonable to plan to grow grain in the space farm at a rate of 850 pounds per acre per day. Plants will grow in sand, vermiculite, styrofoam, or nothing at all provided they are supported and receive nutrients and water. Carolyn and Keith Henson, Tucson agriculturalists, propose to support the plants by means of styrofoam boards. The roots would hang below the boards and it would be possible to spray a nutrient solution onto these roots directly. A significant advantage of this method is that the roots could be harvested for animal feed.

The productivity of wheat and grains can be exceeded by vegetables. The best yields for vegetables commercially grown come from greenhouses in the desert of Abu Dhabi:

vegetables	*lbs. per acre per day*
tomatoes	920
cucumbers	1000
cabbage	530
radishes	560
broccoli	315

Melons would yield about as well as cabbage and potatoes about as much as tomatoes. Potato harvesting would be especially easy with unsupported roots. You could go down the row and pick ripe potatoes as you would pick fruit. Nor do the Abu Dhabi productivities represent the ultimate. By carefully optimizing the growing conditions, these yields might very likely also be doubled.

The colonists can have plenty of grain and vegetables and they can also have fruit. There will be numerous parks and trees throughout the colony—the trees providing shade and making the surroundings pleasant to look at. Trees can have apples or oranges, pears, cherries, plums, or peaches. Even coconuts and bananas will be easy for the venturesome to grow, for the mild, pleasant conditions within the colony will prove suitable for any fruit tree which is desired. Fresh fruit from the parks will be a common item on the colony's menu.

The colonists will also want meat and this poses a problem. Is it feasible to raise meat in the space farm, when space in the farm is at such a premium? The ideal animal must have high productivity. For instance, if you have a herd of cattle, only 20 percent of its mass can be harvested as meat per year. But chickens and rabbits reproduce so fast and grow so rapidly that a herd of either animal can produce five times its initial weight in edible meat over a year.

Chickens or hogs grow rapidly and pound for pound eat much less feed than do cattle. They have often been suggested as being suitable for use as meat animals on a space farm. However, for efficient growth these animals require a diet which is suitable for human consumption. Because of the waste involved in feeding animals a diet which puts them in competition with human beings, it does not seem desirable to choose these animals as the main source of the colony's meat.

Dr. Kenneth Olson, of the University of Arizona and a friend of the Hensons, has proposed raising alfalfa for rabbit feed. Alfalfa gives good yields of protein but has too much fiber for people to accept. The addition of a little salt makes alfalfa suitable as a complete feed for rabbits. Rabbit meat is low in fat and can be cured like ham or made into sausage and liverwurst. It is mild-flavored and can be cooked many ways, even as rabbitburgers.

One square yard will house a doe and her litter. Every two months, a new litter will arrive and that is when the young rabbits will be taken for their meat. Each such doe-and-litter "unit" requires about a dozen square yards of area for its alfalfa. Overall, the production of boneless meat comes to 145 pounds per acre per day. In terms of its protein content, the productivity is as good as that of protein from grain.

This farm produces food for two additional kinds of animals at no additional cost in growing area. Ruminants can convert the waste materials—stems, leaves, and roots from vegetable production—into milk. For instance, tomato vines are up to 24 percent protein. Cucumber vines, melon vines, and cabbage leaves also are valuable feeds. Ruminants could also eat straw or sorghum stalks from the grain production.

There is the question of which ruminant to choose. The two most common milk-producing ruminants are the cow and the goat. Cows weigh ten times as much as goats and eat ten times as much feed. But a cow will produce only four times as much milk as a goat. For a given amount of feed, a goat will produce more than twice as much milk as a cow.

Goats require more care than cows if their milk is to be of high quality. They should not be fed onions, the taste of which will wind up in the milk, and must be kept clean. A dairy goat and her milk will smell bad if there are billy goats around. But the billies can stay on Earth and artificial insemination can be used whenever new goats are needed.

In addition to the forage feeds, such as stems and leaves, goats (as well as cows) require some grain. The space farm can easily raise several times more grain than people will eat as bread or as products made from flour. The excess can be goat feed. Everyone can get two quarts a day of goats' milk, rich in protein and with higher-quality protein as well as more minerals and vitamins than the grain used as feed. Much of this milk may be made into cheese or butter or cream. And the space farm can produce ice cream!

It will also be possible to raise chickens on the farm, feeding them with kitchen waste as well as leftovers from meals and the waste from rabbit butchering. These foods have traditionally been fed to hogs as well as chickens, but egg production supplies food more efficiently than the production of pork or lard. Without adding any area to the farm for growing chicken feed, the wastes will support enough hens to give everyone in the colony three or four eggs per week. The colonists can enjoy cakes, waffles, omelets, and mayonnaise.

To grow enough nutritious, varied food for all 10,000 colonists would require a farm of only 100 acres. Even this is not the ultimate. If the production of grains and forage can be doubled again to 1700 pounds per acre per day—given optimum concentrations of carbon dioxide as well as optimum conditions of light, temperature, humidity, and nutrients—the space farm can be cut down to sixty acres.

The inside of the Stanford torus offers a clear area of 200 acres. In addition, the agricultural areas can be built in levels, with sunlight directed by means of mirrors to each level. The space farm will require only a relatively small share of the colony's space.

There will be an opportunity to expand, to grow additional feed and forage and give even greater variety to the diet. An increase in the number of hens or chickens would be one of the easiest things to do. One of the most popular decisions, along this line, would be to import a small herd of Herefords or other beef cattle and set up a feedlot. (Since feedlots are exceptionally smelly, probably the best place for it would be in the spokes leading inward to the colony hub.) Cattle are rather wasteful at converting feed to beef. They need over twice as much feed as rabbits to produce a pound of edible meat. What is worse, cattle feed includes a lot of corn or grain which can be eaten directly by humans. But people like to eat beef and if the colony can grow the extra feed for cattle, the steers and cows will be very welcome.

There is another popular source of protein which comes from an animal and which may be nearly as productive as the rabbit. This is fish. In a weightless space farm, it may be possible to raise fish without water. On Earth, when a fish is taken from water, gravity makes its gills collapse so that it cannot get oxygen. In weightless space these same fish might easily "swim" through an atmosphere of 100 percent humidity, keeping comfortably moist: hydroponic fish, if you will. In the space farm there will be artificial gravity and this will not be possible. Instead, the fish will grow in ponds at the top levels of the farm, where a quarter million may live. These will be enough to supply everyone with ten one-ounce fillets per week.

In those waters, warm, shallow, rich in phosphates and other nutrients, there is the opportunity to recreate the food chains of the most productive fishing grounds. There will be diatoms, tiny microscopic vegetables, to grow on the minerals, sunlight, and carbon dioxide. These and one-celled algae will be food for the fish. The New Alchemy Institute of Woods Hole, Massachusetts, has developed a "backyard fish farm–greenhouse" which raises crops of fish this way. It grows dense blooms of algae to provide food for herbivorous fish. Tilapia, a herbivore from Africa, has been cultured to edible size in as little as three months.

→

From *Science Year, The World Book Science Annual.* © 1975
Field Enterprises Educational Corporation.

View from inside a colony, showing how one of the mirrors reflects sunlight.

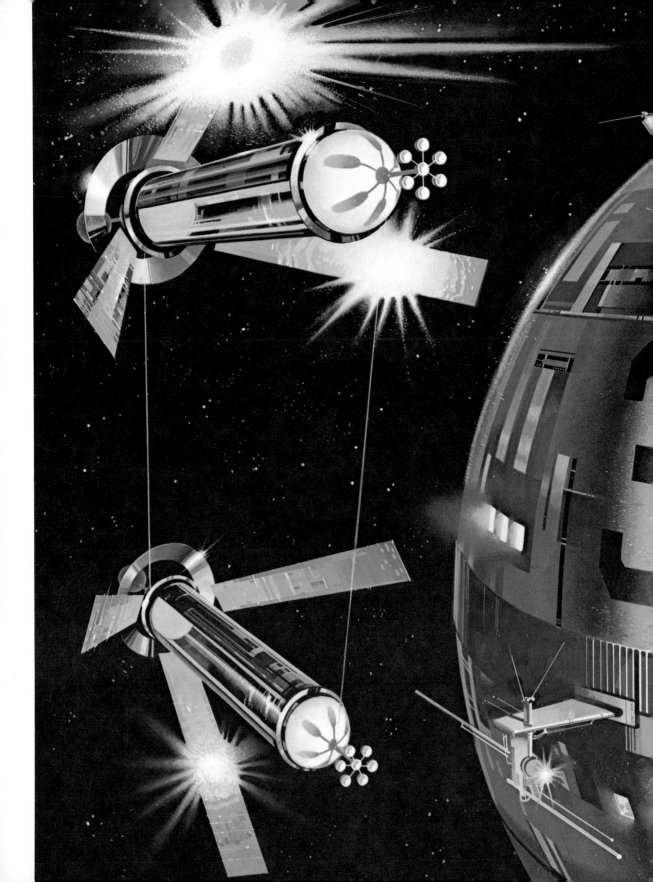

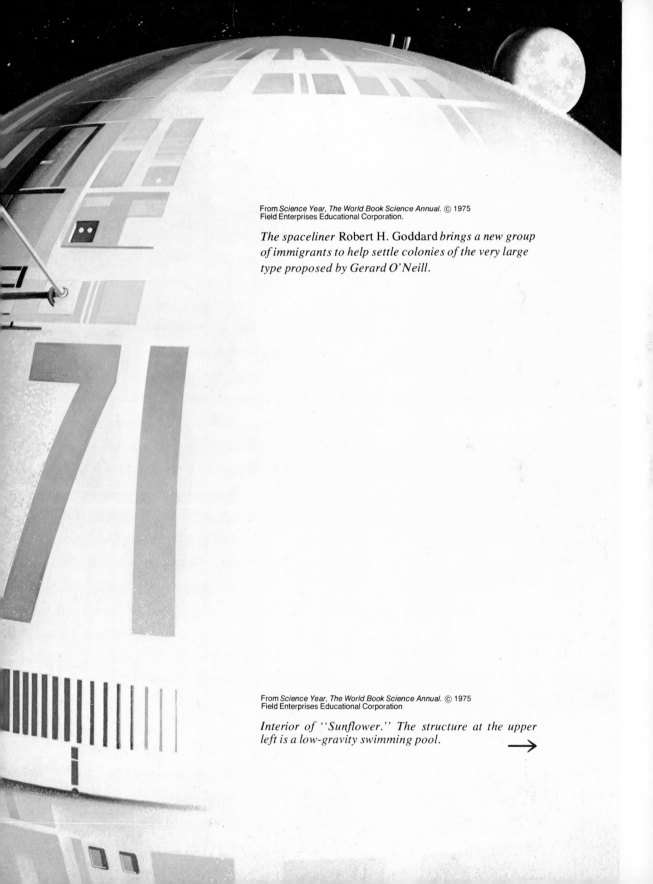

The spaceliner Robert H. Goddard *brings a new group
of immigrants to help settle colonies of the very large
type proposed by Gerard O'Neill.*

*Interior of "Sunflower." The structure at the upper
left is a low-gravity swimming pool.* →

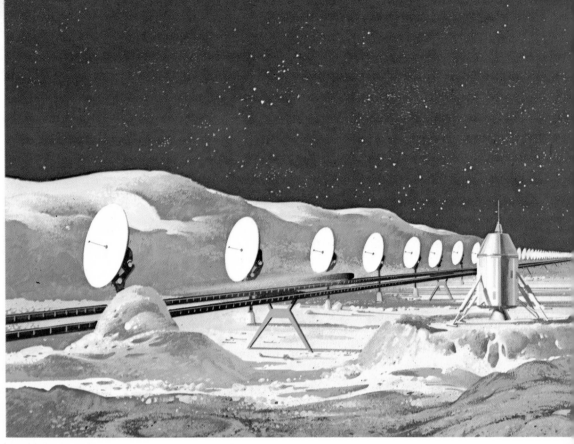

From *Science Year, The World Book Science Annual.* © 1975
Field Enterprises Educational Corporation.

The completed lunar base. Rockets such as the one in the foreground have delivered elements of the base, including the housing and agricultural modules at the right. This base supports the operation of the mass-driver, at left, adjacent to a row of communications antennas.

From *Science Year, The World Book Science Annual.* © 1975
Field Enterprises Educational Corporation.

View of the largest colonies, showing the agricultural cylinders.

(*final picture*) Painting by Don Dixon.

Interior of Stanford torus type colony. In the foreground is one of the small towns; in the rear is an agricultural area. The buildings are sturdy structures made of brick, steel, aluminum, and glass, all obtained from lunar materials. Note the rooftop gardens. No space is wasted in the colony.

Terraced levels for space-colony agriculture. At the top level are fish ponds, the effluent from which goes to irrigate the lower levels, on which are grown wheat and soybeans, vegetables, and forage for the animals. (Courtesy NASA)

The white amur, a vegetarian fish prized in China, has grown to over a foot in length in less than a year.

Will the space farm actually produce these yields? On Earth, farm yields vary quite sharply from year to year and sometimes fail entirely. However, there are reasons to believe that a space farm would be more stable, more predictable than an Earth farm.

On Earth variations in yield are due to weather, weeds, insects, rodents, and disease. Weather simply does not apply in space, assuming there are no breakdowns in the systems for temperature and humidity control (and no one opens a window!).

The initial seeds for a space farm can be individually inspected to keep out weeds. If a few weed seeds do get through, the farm area is small enough so they will be spotted and removed. Simple fumigation of shipments from Earth should eliminate the insect problems, but if undesirable insects get in, they can be dealt with by the means used to fight disease, or lizards and frogs can eat the insects. This is the method used at the New Alchemy Institute. Rodents are easier to keep out than insects and this will not be a problem if scientists do not bring in rats and white mice. In the event they do and some escape, a cat may be necessary.

Disease organisms will be harder to keep out or control, and may come in with each new shipment from Earth. The space farm environment will be rich, similar to that of a greenhouse. Molds and viruses have been troublesome in greenhouses, but there are a variety of control methods. Steam sterilization, for instance, might be satisfactory. Molds are particularly troublesome in the high humidity of greenhouses, but in space the humidity can be kept as low as desired. For grain crops, in any case, low humidity is essential. Moreover, the sciences of coping with diseases date back to the time of Pasteur and the space farm will certainly stock vaccines and antibiotics. Bacterial and virus diseases usually attack a narrow range of hosts, sometimes even a single species. The wide variety of crops and animals grown on a space farm will give further insurance against a widespread disease-caused failure of the farm.

Nevertheless, for those convinced of the general perversity of things (Murphy's Law), there are a number of fallback positions. For one thing, the diet available is rather excessive, producing two or three times more protein and somewhat more calories than are actually required. A substantial fraction of the crops could be lost and no one would go hungry. There is a large safety margin in terms of stored food and seeds, food being processed (e.g. cheese being aged), and fish still swimming or meat still hopping.

Even so, imagine a disaster, nature unspecified, that kills every plant on the farm but leaves the people and animals unaffected. This would be similar to an Earth crop failure. What would be done would be to butcher all the animals which could not be fed beyond the next two weeks and freeze or dry the meat. After cleaning up the mess, the farmers would plant the fastest growing seeds available.

The CO_2 content of the air would not reach a problem level for at least two weeks even with nothing growing. With planting, in one week the CO_2 level would be on its way back down. In two weeks, the production of forage for rabbits would be back to normal and in three months the entire farm would be back to normal. Those two weeks of leeway, before CO_2 levels become uncomfortable, would allow time for the ultimate fallback position: the arrival of help from Earth.

In a sense, this ultimate fallback is the way the farm will be started, as well as maintained in the face of slow leaks or losses of material, which will very likely occur. At the outset, it will be necessary to bring from Earth the seeds, the styrofoam boards, and the initial stocks of animals, as well as stores of carbon, hydrogen, and nitrogen. These last will produce CO_2, which the plants will use for growth, as well as water and part of the atmosphere. Oxygen will come from lunar rocks and will be continually available for replenishment. The same is not true of carbon, nitrogen, or hydrogen, or of compounds which incorporate these elements. Some of them can come from the moon as trapped atoms released when fine lunar soil is heated. The initial stores, as well as most of any additional quantities needed, must come from Earth. The freight rates will not be cheap. Even when the space transportation is working with maximum effectiveness, the shipment costs for the hydrogen in a gallon of water still will come to $40.

It is essential to have effective means for reclaiming water and for restoring carbon dioxide to the air for the plants. Wastes must be purified and converted into useful products without producing pollution. Minerals and fertilizers must be carefully recycled. The methods used do not need to be totally effective, for the colony is not the same as a starship which will cruise in self-sufficiency for centuries. The recycling must be complete enough, however, to reduce the need for resupply from Earth to a level compatible with normal trade and commerce. Perhaps it will be common for the space freighters on the outward trip to carry a few bags of fertilizer or briquettes of charcoal, a few bred nanny goats or chickens, or 100 pounds of salt. But these should be only minor items on the manifest.

To reclaim water, there are two good methods and both will probably be used. The simplest is to extract it from the air with a dehumidifier. The moisture-laden air passes over a coolant and gives up some of its heat. When cooled, the air passes through a screen made of materials which are hydrophilic (water-loving). The moisture condenses and is extracted, to be pumped to a tank. As with many air conditioners, the dehumidifiers will be tied in with the colony's temperature-control systems. The heated coolant will pass through a heat exchanger, the excess heat then being taken to the central radiator.

Most of the moisture in the colony's atmosphere will come from the plants as water evaporates from their leaves in a process called transpiration. Most of the dehumidifiers will be in the farm areas. The temperature and humidity will be controlled by adjusting the temperature of the coolant and the rate at which air passes through the dehumidifiers. In particular, some areas will have to be enclosed and kept at very low humidity—the crop storage areas for instance.

The water from the dehumidifiers will be available for drinking, cooking, and taking showers or baths. It will be a pure water with no chlorine or other chemicals, and it will be free of the dissolved salts which make water hard. It will be like rainwater and it will percolate through no ground strata, need no water softeners; it will simply be condensed from the air and piped to people's homes.

There is another source of water, which will furnish a product well suited for crops and animals—water reclaimed from wastes. The problem of waste treatment requires a solution which not only gets rid of the wastes, but which turns them into useful products. The usual processes used in Earthside communities, such as biological degradation or incineration, are unsuitable for the colony. These processes either produce pollution, or are incomplete since they produce a very messy sludge which must be disposed of.

There is a process which has neither of these defects—the wet oxidation or Zimmerman process. Wastes are heated with oxygen to a temperature of 500° at 100 times normal atmospheric pressure. They are cooked under these conditions for an hour and a half. What comes out is high-quality water containing ammonia and fine phosphate ash, as well as an effluent gas rich in carbon dioxide but free of oxides of nitrogen, sulfur, and phosphorus. Both the effluent gas and the water will be sterile.

The gas will be fed to the space farm to maintain desired levels of CO_2. The water will serve for growing food. The ash and other solids can be filtered, then mixed in with animal feed or with water for the plants as a source of minerals. If more than 2 percent of the initial waste is combustible solids, the process will be self-sustaining. Like an incinerator, it will operate continuously without an outside source of heat, the heat released in the combustion being sufficient to keep the process going.

All the water from the waste processing, as well as much of the water from the dehumidifiers, will first be fed into the fish ponds. From there, the water will be screened to remove fish waste, then pumped to the agricultural areas. What water is not used for the plants will flush human and animal waste to waste processing. In a continuous cycle, water will circulate throughout the colony, being cleaned and purified when necessary and serving all purposes.

The plants will also remove carbon dioxide from the atmosphere. They are entirely adequate for this, and there will be no need for anything but photosynthesis to produce and recycle the oxygen for the colony's atmosphere.

The atmosphere as a whole will follow the principle, "less is enough." It will have enough pressure and there will be enough of the different constituent gases to meet all biological needs. Its overall pressure will be at one-half that of Earth's atmosphere. This will not only make it easier to transport the atmospheric nitrogen from Earth, but also make the Stanford torus much lighter and easier to construct. A thicker atmosphere would need thicker walls for the torus in order to contain it.

It will contain as much oxygen as Earth's atmosphere at sea level. On Earth this much oxygen represents 21 percent of the total; in the colony, it will be over 40 percent. There will be only half as much nitrogen in the colony atmosphere as there is in Earth's. There is no clear reason why humans or animals need to have any nitrogen at all in the air they breathe, but the nitrogen will serve some useful purposes. It will make breathing easier and will reduce fire hazards. Nitrogen-fixing bacteria will help provide it to plants, which need nitrogen to grow.

About 2 percent of the atmosphere will be water vapor to supply humidity. The colonists probably will be most comfortable at a temperature of about 72° and a humidity of 40 percent. Less than 1 percent of the atmosphere will be carbon dioxide and pollution will be controlled to very low levels.

Pollution control methods are highly developed from research in submarines. There will be no autos in the colony and industry will have its own recycling arrangements. But there will be unpleasant or polluting chemicals released into the air formed during cooking or operating backyard barbecues, not to mention the smelly gases exuded from wastes. However, these will be easy to remove. They can be concentrated by being adsorbed on activated charcoal, then sent through waste treatment. Or they can be disposed of directly by passing atmospheric air through a catalytic burner. Some substances, such as mercury and Freon, cannot be disposed of this way. There will be few reasons to bring mercury into the colony, but the exclusion of Freon means that aerosol cans will have to use nitrogen or carbon dioxide. The Navy has a rule not to bring anything into a submarine which is harmful after passing through a catalytic burner and this will be a good rule for the colony. As far as monitoring pollutants is concerned, there are instruments such as mass spectrometers and gas chromatographs which can detect incredibly low concentrations.

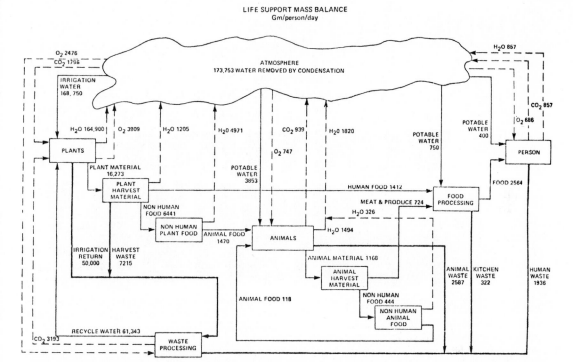

LIFE SUPPORT MASS BALANCE
Gm/person/day

Closed-cycle ecosystem for the colony. Following the analysis of Harry Jebens, such a diagram shows the mass flows between different parts of the colony system. Using mathematical methods, such a diagram then can be used to find whether or not the system is in balance, and whether it is stable, as opposed to being easily upset. (Courtesy NASA)

The colony can be a completely closed ecosystem. The atmosphere can stay pollution-free through two entirely different means. There are also two independent methods to recover water. It should be possible to provide plenty of varied, nutritious food for all with an ample margin against crop failure. The plants, animals, and fish will represent three distinct food sources, each largely independent of the other, providing not only extra variety but an extra margin of safety.

Many people have said the space colony will be a cold, artificial place where people will be cut off from nature. But the colonists may have more involvement with growing things than they would have in a city on the earth. Many of them might well spend several hours a week in the space farm, then go home to eat the meat or cook potatoes and vegetables which they have grown with their own hands.

There is one job which probably will prove too tedious to attract volunteers. This is the hand-pollination of vegetables. For this, the farm should include several hives of docile bees bred without stings or selected as particularly slow to get angry. These bees then will complete the space farm as "a land of milk and honey."

Chapter 10

VENTURA HIGHWAY REVISITED

One of the really pretty parts of southern California is the eastern San Fernando Valley. The towns of the eastern valley—Encino, Tarzana, Canoga Park—represent rather attractive exercises in community design. There are a few high-rise banks or hotels, but these stand alone and not clustered as in downtown Los Angeles. Most of the buildings are small affairs of two or three stories at most, set somewhat closely together yet interspersed with numerous trees. A few blocks to the south of the highway, there are residential areas. The homes are often of open-frame designs with lawns and gardens. The streets wind through these areas, which run up the hillsides.

All this is rather like the architecture and design which can be imported into a space colony. There, too, there can be an emphasis on closely spaced low-rises, on having plenty of trees and greenery, on simple, open designs for homes. While this analogy should not be pushed too far, there is no doubt that such communities can inspire our planning for the towns and villages of space. From the start, we can seek to pattern them after the most pleasant and attractive Earthside locales.

This represents a departure from some of the usual science-fiction scenarios of space communities. There, discussions of immense mega-monstrosities 200 stories high like the buildings in the Otis Elevator magazine ads are common. Or Buckminster Fuller-type "machines for living" which no one will ever be inspired to surround with a white picket fence. In the colony none of these approaches will be useful.

The landscaping and architecture there will be wholly artificial; there will be no natural

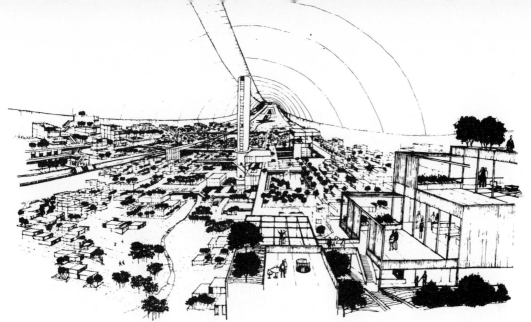

Residential district inside the colony. (Courtesy Pat Hill)

hills or lakes to guide the design. Everything in the colony will be ultimately derived from materials brought from Earth or the moon and will represent things not naturally found in space. The architect's main problem will be to cope with this artificiality, to build within it a community which people will enjoy living in.

The designs must have integrity; they must be valid statements of what it is to live in space. The integrity will not be easy to achieve. There will be architects of the "Tree-Grows-in-Brooklyn" school, who would build the colony as an extraterrestrial version of some of the big apartment complexes of our cities. This school of architecture would emphasize efficiency, uniformity, and regimentation—and then, just to show that all is not concrete and clay, they would give us a tree. It might be a real spreading oak or an oak made of styrofoam or a stylized one made from titanium mined on the moon, or perhaps even a projected three-dimensional image of an oak.

Other architectural concepts could easily lead to similar travesties in a space colony. There might be a simulation or replica of some desirable Earth environment such as a seacoast village or a town high in the mountains. But this would be transparently artificial, out of context. The seacoast would lack a sea, and the colonists would not be living lives appropriate to such seaside communities. The mountains or hills, being merely artificial, would irritate people who have to climb them and who would prefer grades less steep.

Any attempt to model the colony architecture too closely after an existing Earthside community must inevitably degenerate into a type of play-acting with people living in what is essentially scenery for a play. The colony town names may be taken from Earthside and we may see names like Ventura or Monterey or Vail, but any attempt to recreate the real Vail or Monterey will only fall flat.

No, the settlers of space will do what people in all ages have done when building new cities. They will rely on native materials chosen for their similarity to those with which they are familiar. They will develop forms influenced by those of their former homes and towns, yet shaped and structured by the characteristics of their new lands.

There is one science-fiction cliché which we can immediately dismiss. This is that the homes will use plastic, perhaps even be built entirely of plastic. Plastics are made from hydrocarbons such as oil and these will be quite costly in the colony. The precious stores of carbon will cycle through the plants and food, not be locked up in the walls of a home unless there is a clear need (for example, as insulation for wiring). For the same reason, wood may be a rarity even if it is locally grown.

The builders will rely on materials readily available. These include the metals aluminum, titanium, and iron or steel. Also there will be plenty of glass, including fiberglass for insulation or soundproofing. Other building materials will also be available from the lunar materials brought to the colony.

For instance, many lunar soils are chemically similar to clay. It should be possible to mix them with water, make bricks, and fire them in a solar furnace. The resulting bricks will usually be gray or chalky in color rather than being red or light brown. Bricks could also be made of cast basalt—a material used in France for tiles, plumbing pipes, and the like—from lunar basalt melted in the same solar furnace. Quicklime, obtained from lunar plagioclase and mixed with water and dry lunar soil, can be used for cement. It is quite common on the moon, nearly as abundant as aluminum. There would be concrete for sidewalks or walls, concrete blocks for foundations, and thin, watery cement to serve as stucco for houses.

Steel, aluminum, brick, glass, concrete. Of these would the towns be built, these solid substantial materials of integrity. A home in the colony would be durable and lasting, well worth holding.

It would hardly be desirable to arrange for the homes to be built as in Earthside developments, where a small army of workers erects them from piles of planks and two-by-fours. They should be constructed with a minimum of human labor, perhaps even by only the people themselves who are to move in. This means they will be built from modular components, from standardized sections and elements which can be turned out by the thousands in a specialized area of the construction sphere.

Modularity does not mean uniformity, but represents a seldom used approach to constructing a community. All towns, however diverse, are ultimately constructed of very large numbers of nearly identical bricks, concrete blocks, two-by-fours, and wiring or plumbing fixtures. This represents one extreme of modularity. The other extreme is represented by suburban developments where all homes are identical except for the shape of the TV antennas. There, the modularity is at the level of entire houses.

In the colony the modularity would be at an intermediate level: wall panels, windows, roof sections and the like. Though uniform in size, these could be combined in a very large number of variations. The architects would be working with standardized elements, seeking to develop a larger number of floor plans and home designs which would be built from them.

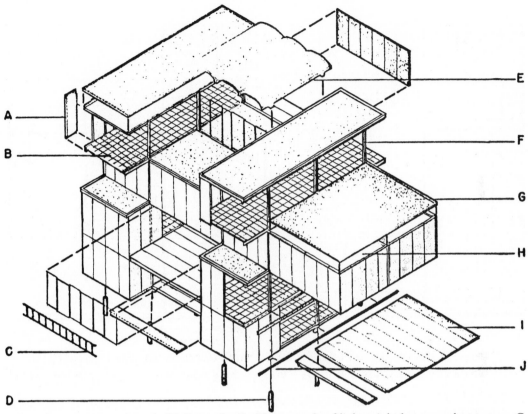

Colony home construction. A. Wall panels. B. Floor panels of lightweight honeycomb structure. C. Shades and railings added in place. D. Structural supports for the frame. E. Roofing kit, providing a clear, colored, or opaque ceiling. F. Structural frame stacks four stories high. G. Roof panels for use where a roof is to serve as a surface for walking. H. Ceiling panels. I. Spanning planks or beams. J. Structural beams. (Courtesy Pat Hill)

The resulting structures would be light and open with the major structural shapes defined by steel girders and aluminum beams. The walls would bear no heavy loads and could be interchanged when desired with large floor-to-ceiling windows. A popular design will likely be a corner living room enclosed in glass on two sides. Kitchens and bathrooms will be assembled at the construction sphere as complete units with refrigerators and dishwashers and ovens or toilets, bathtubs, and sinks installed in place. The homebuilder would simply hook up the plumbing and electrical connections to local mains in the community.

There will be studio apartments as well as one-, two-, three- and four-bedroom homes with up to 2000 square feet of interior area. The larger ones could be of two-story construction. In addition there would be flights of stairs for reaching adjacent homes as well as the main ground level.

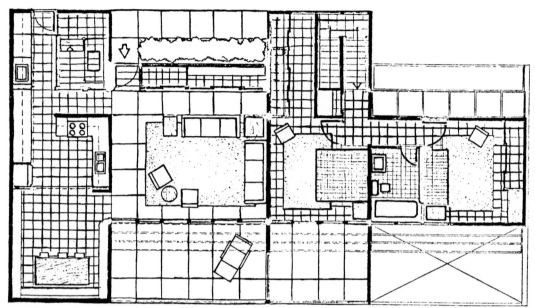

Floor plan for a two-bedroom home. (Courtesy Pat Hill)

During the 1975 Summer Study, Pat Hill, an architect from San Luis Obispo, California, devoted attention to layouts and delineations of the colony interior. He describes a likely home as follows:

> *Although small, the apartment is completely furnished in a compact, convenient, and attractive way. The furniture and the decorations are made of aluminum and ceramics; fabrics are made from lunar silica. Perhaps a homesick previous tenant put up over the sofa the large picture of a full Earth filling a black sky. The ubiquitous use of ceramics and aluminum is a constant reminder that all organic chemicals must come from earth. . . . It will take a while to become accustomed to the almost complete absence of wood.*

The homes or apartments will not be simple duplicates of existing Earthside designs. The conditions of the colony will dictate some distinct changes. In the kitchen, microwave ovens will be popular because they cook faster and use less electricity and can be built around elements similar to the Amplitrons used in the power satellites. Few homes may have their own washers and dryers, since it will be easy and convenient to provide laundromats serving clusters of houses.

Some draperies and carpets may come from Earth along with the rest of the colonists' household goods, but what is made in the colony will be of Beta-cloth. Beta-cloth is a fabric woven of glass fiber, originally used by the Apollo astronauts because it is fireproof. It is not rough or abrasive and will not scratch the skin but is grainy and a bit stiff. It is somewhat like denim fabric for Levis, not so stiff as canvas, perhaps like thick starched cheesecloth. It would never be useful for undergarments but would do quite nicely for upholstery and draperies. It also could be tufted on a thick mat for carpeting.

In one respect, it is quite unlike ordinary fabrics. It will not soak up dye. Any dye applied will quickly rub or wear off. However, Beta-cloth may be woven of colored glass. Since the Middle Ages, artisans have known how to tint or stain glass, to produce rich and varied blues and reds, purples and greens or browns. The art consists of adding traces of metals to the molten glass: iron for brown, cobalt for a rich blue.

The staining of glass may be one of the colony's art forms with stained translucent glass being incorporated into walls and parts of ceilings. Curtains and draperies of stained Beta-cloth may be beautifully iridescent, glowing in the morning sunlight. The same glowing, prismatic colors can come from the floor underneath by laying Beta-cloth carpeting on top of a polished aluminum floor covering. What's more, housewives will appreciate the freedom from stains of Beta-cloth since spills or spots will wipe off with a damp sponge.

The colony homes will also be brilliant with flowers. The warm mild climate will be ideal for growing the lovely flowers of the tropics: frangipani and hibiscus, croton and maile and bright orchids. They will need frequent watering, but the splendor of their growth will repay the effort involved. Other people will prefer less exotic flowers: geraniums or sunflowers or morning glory, lilacs or buttercups—or, for contrast, a cactus growing in a rock garden in the front yard. To set off the flowers and the colored glass, the stucco or aluminum walls of the colony can be painted with light pastel tints.

Individual homes would be characteristically homes of the colony. Their materials, their construction, their decoration all would be statements of what it is to live in space. They could be open and colorful as well as comfortable to live in. But what of the community as a whole? Would it possess similarly attractive design?

The community environment must provide both complexity and ambiguity. That is, it must not be so complete as to seem sterile, capable of being taken in with a single glance. It should be possible to stand on a balcony and see a profusion of detail, much more than can be absorbed immediately, so that a look time and again would still discover more new things. Colonists will need large spaces and smaller ones, private and community areas, long vistas and short ones. The Stanford torus design will aid this sense of incompleteness or ambiguity, since at any time about five-sixths of the colony will not be visible in a single scan but be out of sight around the curve. The great variety of flowers and trees also will lessen the sense of artificiality.

For all this, the community planner will be faced with one simple fact: The available land per person will be 510 square feet, about the same as a living room 23 feet square.

This is the area of the main level of the torus, shared among 10,000 people. If someone builds a two-story house 23 feet square, he has doubled his living space but used up all his allocation of land. If there is a multi-story high-rise, then people can enjoy large living areas indeed, yet each will have the equivalent of only a small amount of land. Land in the colony, in this sense, is the same as area or acreage of land as it is usually sold. The colony will be in much the same situation as many city centers where land is so scarce and expensive that only a high-rise will do.

For comparison, the available 510 square feet per person is similar to the 415 per person of Manhattan and the 435 of Rome. There are, by contrast, 1783 per person in San Francisco and 5460 in Columbia, Maryland, a town planned from the start as a pleasant and attractive place to live. There are worse places. Hong Kong offers about 100 square feet per person. And the award-winning architect Paolo Soleri, whom Stewart Brand describes as an "urban visionary," has designed a high-rise community which allows 164.

To convince ourselves that such a crowded community need not be unpleasant, we must look for some Earthside communities which are attractive yet low in land area per person. Such towns do exist and from them we can learn some of the ways to make the colony a desirable place to live.

Some outstanding examples are the old villages of France. There are places like Saint Paul and Vence in the south of France, where people have lived for hundreds of years in population densities which would give even a space colonist claustrophobia. Saint Paul has 293 square feet per person, Vence has 500. How do you deal with this shortage of space so the inhabitants will not feel oppressively crowded? The answer is you hide it by breaking up the community into small clusters of houses, each with a courtyard and a fountain or group of trees. Then people will usually be aware only of the few houses and people immediately at hand. This is entirely distinct from living in a high-rise, looking out the window at another high-rise and knowing only that there are more people about you than you would care to meet. Instead, such small clusters foster a sense of community. You would look out the window and see the homes of your neighbors, if not of your friends.

Both in Saint Paul and Vence, the humanness of the scale is accentuated by the absence of cars. The experience of walking through a small town becomes entirely different when there are no automobiles. Without them, you can think your own thoughts as you go or notice flowers or lawns. With them, if you are not yourself preoccupied by being behind the wheel, you are very aware of cars and the need to watch out for them. Only in a very spacious town, like Columbia, Maryland, can cars and pedestrians live side by side in some semblance of harmony. In more crowded places, one or the other must predominate and in most locales the decision has favored the automobile.

In favoring the pedestrian, the colony would resemble Saint Paul or Vence. Distances will be short—no part of the colony will be more than two miles from any other part. And there is a simple act of community design which will reduce distances further, enhance diversity more strongly.

There are six spokes to the Stanford torus leading to the hub, and half the periphery of the colony is given over to agriculture. It would be easy to divide the periphery into six approximately equal segments, three for agriculture alternating with three for people to live in. Each of the populated segments would be a little town with the population a bit over three thousand, about five-eighths of a mile long and one-thirteenth mile wide, centered on one of the spokes.

The spokes with their elevators would be the main transportation links between parts of the colony. To go from Ventura to Vail, you would walk to the base of the Ventura spoke. It would not appear as a spoke, of course, but as a single high-rise, twenty stories tall. It would function as a high-rise with space for offices and for a few apartments. An elevator would arrive, having descended the outside of the spoke in a glass-enclosed tube. You enter the elevator along with several other people and it soon ascends. On the way up, the large window in the elevator and the glass in the tube permit a view of Ventura as it falls away. Then you are through the ceiling, traveling through space. The sun glints brightly on the adjacent spoke on which other elevators run and there is the immense curve of the colony as a whole, seen more clearly as you approach the hub.

Once at the outer edge of the hub, the door opens and you debark into a tunnel running around the periphery. Your weight has been dropping all the way up until now it is only one-sixteenth of normal. The tunnel has thoughtfully been provided with handrails and a padded ceiling to keep people from being hurt if they launch themselves upward. Since you are heading for Vail, you follow the tunnel past the first elevator station (to the Kew Gardens agricultural sector, lying between Ventura and Vail) and on to the second station, about 400 feet from where you got off. Soon the next elevator comes and you are on your way to Vail.

These elevators will be in addition to others within the spokes and used by people working there. (Among the elevators will be large freight carriers to carry payloads from incoming spacecraft down to the level of the colony.) But the spokes and their elevators will be unusable during solar storms when cosmic-ray levels mount to dangerous heights. Moreover, not everyone will want the changing g-levels or the inconvenience of walking through the crowded tunnel at the hub (perhaps with small children or bags and parcels). The colony will need a separate and entirely distinctive transportation system, a system that can take advantage of the fact that below the main level of the colony there is ample enclosed space. It will be easy to build electric subway cars, running between stops. These will be small, light railed vehicles to carry passengers around a periphery of a few miles. It will be desirable to have them controlled by their passengers to some degree, like an elevator. There should be buttons to push in order to open or close doors, or stop. Indeed, such cars could be designed as horizontal elevators with station signs as in a subway. The idea is there should be some involvement by people in their own transportation, even if it is simply that of pushing buttons and reading signs.

By relying on elevators and subwaylike people-movers (or perhaps the science-fiction cliché, the monorail), the colony's transportation needs can be met while keeping the living areas free of traffic. The rest is easy: walkways for pedestrians along with bikeways. Much of a community's structure depends on its transportation and these arrangements will make it much easier to design Ventura, Vail, and Monterey as human civilized places built around the needs of individuals.

The homes can hardly stand isolated, each on its own lot as in an Earthside town; yet it is important to give everyone something of the feeling of being a freeholder. By terracing the apartments, setting one atop the other to give everyone a patch of open space outside the front door, individuality is achieved. An outstanding example of a similar design is the apartment complex known as Habitat, designed by Moshe Safdie for the 1967 Montreal World's Fair.

In Habitat, one man's roof is another man's garden. The terracing is arranged so that each apartment door opens onto a neighbor's roof, which is planted with flowers or grass. Significantly, in a city with no lack of the residential areas we are familiar with, Habitat is one of the most popular of Montreal's apartment complexes and is regarded as a particularly desirable place to live.

This style of community planning would be easy to achieve in the colony. Lunar soil will easily support the growth of grass or flowers. It was one of the surprises early in the Apollo flights that plants were found to grow very well in it. With beams for structural support and fiberglass for soundproofing, it will not be difficult to put a lawn on top of the living room.

Scattered throughout the towns will be public parks and gardens as well as small lakes. These will not merely be ornamental. The parks will be the places for fruit trees, each park with a different type of tree. They may be known as Peachtree Park, Orange Grove Park and so on. The lakes will serve as reservoirs for the water supply, but can also be homes for families of ducks and a swan or two perhaps.

In addition to flowers, lawns, and parks, there are other aspects of a natural environment which deserve attention. In much larger colonies, it will be possible to have clouds and weather. This will also be true within the Stanford torus, though to a lesser degree. On Earth, we are accustomed to hearing "It will be partly sunny till early afternoon, followed by showers." We think of these statements as predictions. In the colony they would be announcements.

Unfortunately, though the Stanford torus is large, it is not large enough to produce weather in a natural way. Any clouds to be formed would hang low, 100 feet from the ground. They could be produced only by making the whole colony chilly and clammy, as on a foggy day. The only way to make rain would be artificially, sprinkling it from above. While the flowers and grass would enjoy this and the resulting rainbows would be pretty, the people would more likely look on it with distaste. Some people enjoy walking in the rain, but a walk fully clothed through what such sprinklers would actually be—immense shower stalls—might not be enjoyable.

In the colony there would be plenty of small-scale sprinklers and hoses. It would be a familiar sight to see streams of water in the gardens, spraying back and forth. But since the climate will be entirely controllable, the colonists would resent any changes they did not agree with. There would be none of Mark Twain's fatalism: "Everyone talks about the weather, but nobody does anything about it." Some people from Earth would regard the colony environment as entirely too artificial for their tastes: no clouds, rain, storms, or changes of season.

The colony will not be a paradise. It will be comfortable and attractive, but it will still be a frontier community working hard to build power satellites. Its population will be too small to sustain a major university, a large medical center, or the profusion of specialized shops and stores which are found in all large cities. The people of the colony will have opportunity to exercise any skills or talents they may have, since it is a characteristic of the frontier that there are many things to do and few people to do them. By the same token, the lack of specialization will tend to limit many of the options which would be enjoyed on Earth.

The limited population would restrict the services available. In practice, it might prove not too serious since it would stimulate a do-it-yourself approach to many activities. Another restriction would come from the need to import many things from Earth. Though the colony would be largely self-sufficient for its major needs, there are items for everyday life which it would have to import—tooth paste and brushes, razor blades, needles, glue or adhesives, and perhaps even soap. As with the early colonists of America, these items would come by ship from the mother country.

Among the most important of these everyday items, which would be very advantageous to produce at the colony, would be cloth and paper. Beta-cloth may be the blue denim of space, serving wherever a rugged and durable fabric is needed, but there will be a need for softer lighter fabrics. Portions of the agricultural areas may be set aside to grow cotton, with the bolls for cloth and the stems and leaves for animal feed. However, most new cloth would be made by recycling old clothes.

The clothes would be shredded and reduced to fluffy piles of fiber, similar to raw cotton. This fiber would be spun into thread and woven on looms into bolts of new fabric. Some of this would be made into such everyday items as socks and underwear. There also would be a limited selection of store-bought shirts, pants, and dresses available from the clothing store. Those who want particular styles or who seek especially distinctive or attractive attire, will have a simple way to get it. The clothing store will stock a large assortment of patterns as well as sewing goods. The would-be lady of fashion can make her own dress or look around to see who in town can be a tailor or seamstress. The community bulletin boards will often carry announcements by people seeking or offering these services.

Paper will come mainly by recycling. The old paper would be shredded and reduced to pulp, and made into new. There would not be the opportunity to offer the large range of weights and finishes of an Earthside paper store, but there are at least two grades which could easily be made. There would be a heavy and rough kraft paper for bags and corrugated cardboard in boxes. There would also be a white bond paper for writing and printing.

Downtown in a colony business district. (Courtesy Pat Hill)

The kraft paper bags will be the common means of marketing agricultural produce. The colony will provide little opportunity for a glass-and-chrome supermarket with hundreds of brand names packaged in plastic or enclosed in glass on the shelves. There will be something like a bunch of farmers' markets with open-air stalls, bins full of fruits and vegetables. Some of the stalls will be refrigerated for meat and dairy products. There will be bins of granola-type cereal sold by the pound, cookies and cakes made by enterprising housewives, jars of jam and preserves, eggs, and occasionally whole frying chickens.

All the food will be fresh, coming directly from the agricultural sectors a few hundred yards away. None of the food will spoil in transit or need to be laced with preservatives to retard such spoilage. For all this there are lots of specialty items which the people will have to regard as luxuries available only from Earth.

As for the bond paper, this will be associated with what in the colony will pass for newspapers and magazines. As for newspapers, the colony will quickly start printing its own. But it will be quite impossible to provide magazines from Earth by subscription to individual people. Instead a few magazine copies will be sent to libraries, to be used in printing the colony editions. There will be print shops with fast Xerox machines to make the copies of periodicals for the subscribers. These copies will constitute the space colony editions of the periodicals, printed by arrangement with the appropriate Earthside offices. People will still be able to get *Time* and *National Geographic* (the color-reproducing copiers will be a boon to subscribers to the latter), but they will resemble stapled Xerox reports rather than magazines as we ordinarily receive them.

The matter of magazine subscriptions is closely related to that of mail, and of other information from Earth. Much of the letter mail we receive is either bills or junk mail, neither of which should be a problem in the colony. Junk mail can be kept out simply be requiring all space-bound mail to be first class or air mail. Bills will be handled electronically, as part of the banking and investment services.

Letter mail for the colony, if it were to pay its way, would cost about the same as the rates for the 1861 Pony Express: five dollars per half-ounce. There will be some sort of subsidy with the Space Industrialization Administration relieving at least that portion of the fiscal burden of the Postal Service. In addition there will be telex links and much of the colonists' mail will be printouts from the radio receiver.

Mail from Earth will be extremely important to the morale of the space colonists, for they will have few opportunities to visit their former home. The whole colonization effort will be directed to taking people and equipment out, not to bringing them back.

For the outward trips, both to the moon and to the colony, there will be a well-developed system. Heavy lift launchers, built with a winged and recoverable variation of the old Saturn V first stage, along with a hydrogen-fueled upper stage, will put 200 tons into low orbit. There will be a large space-based rocket—to carry the cargo onward. This rocket will shuttle routinely between low Earth orbit 300 miles up and deep space.

The space-based rocket can carry people to the low orbit, which is 99.9 percent of the distance to Earth, from either the moon or the colony. But there may be no regularly scheduled return service to cover that last 0.1 percent.

For many years there may be only one way to get down: the space shuttle. The second stages of the heavy lift launchers will not be recoverable or reusable; their engines and instruments will be packaged for re-entry, but their main structures will not. The problem arises from this: Space shuttle flights will be common, but opportunities for return by shuttle may be rare.

Some people might get down by a kind of space hitchhiking. There will be a fairly constant shuttle traffic to orbit and back in the routine business of launching and maintaining the world's communications satellites. Occassionally there will be a spacelab flight, and the shuttle will stay in orbit up to a month. On such flights the shuttle often may be in an orbit which the space-based rocket can rendezvous with. A returning colonist could go to the low orbit and hitch a ride on the shuttle if there is room in its crew compartment to rig a jump-seat for the return.

There will be regular flights from the moon and colony to Earth. But these may be reserved for senior administrators, lunar work crews and other people who will be in space on temporary tours of duty. Additional colonists might be accommodated often enough, but this can be only on a standby, space-available basis.

The space colony will be planned on the assumption that most colonists will want to stay and make space their highest home. In general the system will rely on the colonists' being highly motivated and fully aware of their commitment to leave Earth for good.

Will the colonists be homesick for Earth? Certainly the scenic beauties of our planet will be missing in space. There will be no oceans, no mountains or sweeping plains, and many colonists will regret their lack. But there will be a milieu, a natural environment far vaster and grander than any the Earth can provide.

Many of us know what it is to be alone in the open desert beneath a starry sky or on a ship in the midst of the midnight sea. These are only a small indication of the true majesty of space as the colonists will know it. The unutterable vastness of space will shape the lives of the people there. It will provide themes for their art, it will be a source for their literature, it will influence their religion, it will work its way upon their outlook and thought. It may be centuries before we will comprehend the implications of a simple observation: The space colonists will be a people of space.

Chapter 11

WHAT'S TO DO ON SATURDAY NIGHT?

Up to now we have looked at how the colony will be built, how the colonists will live, what kind of work they will do, and why it will be important. But the colonists will be no cadre of grimly self-sacrificing robots. They will be ordinary human beings. What will they do for recreation? When the work is done and it's time to relax and take it easy, what will they do?

An obvious activity, enjoyed the world over and beyond, is sex. And, in particular, there will be sex in zero g.

Thus far the all-male crews of straight-arrow astronauts have had little opportunity to sample the delights of this. But Arthur Clarke, who has foreseen so much of importance in space developments, has been here too. "This much we can predict: Weightlessness will bring new forms of erotica. And about time, too."

One way to enjoy such zero-g delight will be in a space Chevy van. This will be a popular recreational vehicle for sightseeing or cruising in nearby space. It will be a small self-contained spacecraft with a roomy interior, suitable for decorating with comfortable rugs or waterbed mattresses, and with large wraparound windows. It will be equipped with a small rocket motor, a reserve of propellant (alcohol, most likely), a radar, and a simple inertial guidance system. The latter will determine whether the van is on a collision course with some object in space such as a powersat, or whether the occupants are approaching the point where there will not be enough fuel to get back. If either is the case, the onboard computer will sound a warning buzzer—loudly, to be sure of getting the occupants' attention.

These vans will also serve for family outings or excursions. People will want to go out a few miles and see the colony from a distance. Or they will want to go off farther and try the sensation of being lost in space, of being surrounded on all sides by that starry vastness. When there are other colonies nearby, these vans will serve for visits, for travel to and from them. A whole subculture may grow up around these vans. Hot-rodders will buzz in and out around the colony spokes, try to match speed with the colony in its rotation, or fly alongside the elevators on the outside of the spokes. Any incoming spaceship of note, or any major new structure being assembled at the colony, will find its retinue of celestial sidewalk superintendants standing off at an appropriate distance (perhaps at an inappropriate distance) in their vans.

When these vans return to dock at the central hub, the occupants will be close to another attraction—the low-gravity swimming pool. At one-twentieth or, more likely, one-fiftieth normal gravity, the water will certainly stay in place in its cylindrical pool. Also the human body has just about the same density as water. So when swimming under water people will find that, just as on Earth, the main forces on them are from their own swimming and from the drag of the water. But in few other respects will the colony's swimming pools (which double as the local reservoir) resemble Earthside pools.

They will not be flat but will curve around in a circle. You can stand on the tile deck beside the pool and look around. Nearby everything will look pretty normal. But as you raise your gaze you will see the pool curving upward, then arching overhead. Directly above people will be shouting and splashing in the water, and it will be easy to spot the newly arrived Earthsiders by their amazed reactions to this.

What also will take getting used to will be the slow measured motion of the water and the people in it. Waves and ripples in an Earthside swimming pool usually are small and travel rapidly across the surface. When someone jumps in from the high diving board there is a splash, a spout of water, a wave, and then the usual smooth pool surface. But the low gravity in the colony pool means that waves will be much higher. They will be like waves at an ocean beach, yet will travel quite slowly. When someone jumps in, he will make a noticeable hole in the water which will take a second or so to fill up. You can sit on the water in an inner tube or rubber raft, and the waves lifting you will make you think of a slow-motion movie of a ship-wrecked sailor in the middle of the ocean.

Water fights will be great fun, carried out at long distances. A double handful or pailful of water will more or less hold together under its surface tension, forming a glistening blob which squirms and wiggles in its flight. But there will be rules against too much splashing about; it would fill the air with drops of water which would take a while to settle and would disturb other people's fun.

Diving boards will provide many an opportunity to show grace and skill. The common

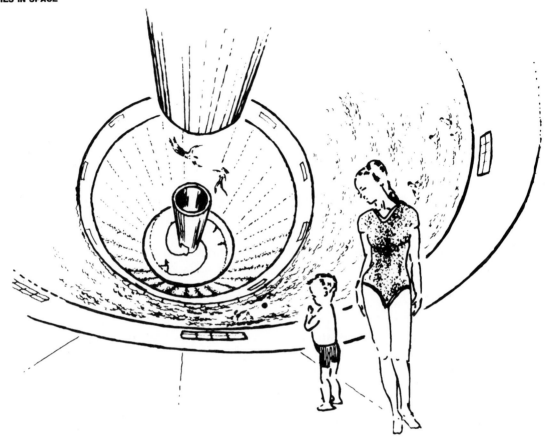

Low-gravity swimming pool. (Courtesy Don Dixon)

garden-variety dive will be a breathtaking slow-motion arc as the diver rises dozens of feet above the board then slowly turns and glides into the water. Like a bird in flight, he will be seen to make small motions with his arms or feet to adjust his orientation. But this is only the beginning. Some people will set themselves spinning when they leave the board and float into the air tumbling so rapidly that they will not know where they're going. Several seconds later, their tumbling will end when they hit the water.

Because there will be the pool not only beneath the boards but also directly overhead, some people will try the vertical dive. This consists of bouncing up from the board with sufficient speed to carry through the center of the pool and on to the pool that lies overhead. There will be endless variations on this for the divers with the most skill. Thus the pool will probably have three or four sets of diving boards, and some people will try to leap from one to the other, gaining an extra spring at each bounce.

Swimming in the air will be almost as important as swimming in the water. By adjusting the jump from a board just right, you can enter the region near the pool's center with very slow speed. Then, by appropriate kicking and waving, you can bring yourself to a halt and sit there, in zero g, while the pool and the colony rotate about you. Little flapping motions will take you hither and yon. Some inexperienced people will find it hard to get back down to the water, so the lifeguards will occasionally have to pull people down from the air. But people will quickly learn the secret of getting down: sit tight, let the rotating air within the periphery of the pool start you moving again with the colony, and centrifugal force will do the rest.

Some people may try to make their own private swimming pools. They will leap into the central zone of the pool, carrying pails full of water. These they will dump out into an accumulation of water at the center, floating in zero g. Instead of making sand castles on the beach, swimmers may try to make water spheres in the air. Then they will try to dive in by means of the vertical dive. Or they may make a big globule and push it so that it will leave the central area and drift back to the pool, resembling a meteoroid made of water. Perhaps someone will even try to put himself in the middle of such a globule with his arms and legs and, hopefully, head sticking out.

A popular sport will be the walk-on-water game, in which you slap the water with the soles of your feet to stay on top of it. But you will have to be careful not to trip over a wave. A good way to get on top of the water will be to swim upward like a dolphin, letting your momentum carry you clear of the water and possibly quite a ways up before you settle back.

Also there will be the flying fish game, in which you slap at the water with the palms of your hands and with flippers on your feet. This lets you skim across the surface at an altitude of a few inches. If you are traveling in the direction opposite to the colony rotation, and get up enough speed, you will soon find yourself weightless, your motion canceling out the motion of the colony which provides weight. You then can become a real flying fish and soar up into the air till you lose speed and the rotation of the air takes you back to the water again.

For those who prefer drier types of sport, it will be easy to arrange an area for human-powered flight. People engaging in this sport will strap on wings, attaching them to their arms and legs, then run forward to get enough speed to take off—that is, to counteract the artificial gravity from the colony rotation. Dipping, gliding, soaring, the human butterflies will cavort amid the inside of the main hub, then swoop gently to a landing. And no doubt someone will advertise waterproof wings, the better with which to play the flying-fish game.

All these will be sports for the colony hub or for weightless space. In the normal gravity of the colony, some people will carry on an Earthside sport, hang gliding.

There are beautifully long tapered kites with wings like seagulls, the SST's or Super SwallowTails. They will soar or hang in the air with even a moderate breeze. These will be the kites most highly prized in the colony, for they will give the best performance. People will start from the tops of the spokes, just below where these pierce the ceiling of glass on the inner periphery of the torus. They then will fly to a landing in a park, fold their kites, and take the elevator to the launching ramp again. Like graceful gulls, they will fill the air with the color of their wings and the ease of their motion, so that people below will look up with astonishment.

Some flyers will use a kingpost motor. This is a small, lightweight engine driving a propeller and mounted on the main vertical strut (the kingpost) of a hang glider. In the colony, such motors will let people skim along just below the glass ceiling or fly completely around the colony interior. With their small buzzing engines, fueled perhaps by alcohol, they will be the colony's dragonflies.

Alcohol, of course, will serve for far more than fuel for the vans and the kingpost motors. With freight rates from Earth at something like $100 for a fifth, there will be little importing of alcoholic beverages from Earth. But there is no need to enforce prohibition. It will merely be true that in this area, as in so many others, the colonists will be on their own. Some enterprising chaps no doubt will acquire control of a supply of grain or other fermentable crops, then proceed to brew whiskey, wine, and beer. Hopefully, not all the hops in the colony will be those of the rabbits. As an alternative to the expense of imports from Earth, this home-grown approach to providing the cup that cheers will be most welcome. Other resourceful entrepreneurs will no doubt find ways to grow tobacco in the colony.

While booze may be scarce in the colony, the same will not be true for first-run movies. Each of the three towns in the colony can have a couple of movie theaters, with the attractions changed every time a new ferry rocket comes up from Earth. These theaters, of course, will be run in keeping with the colony tradition that everything serves several purposes.

The theaters will actually be the community centers. They will be large rooms, with foldup chairs, probably located underneath the main deck of the colony interior. They will serve for holding town meetings or church services. They will be used in their turn by little theater groups, for basketball games or indoor tennis, for dances, weddings, or big parties.

No community is complete without its restaurants. Except for being buffet style, they will be quite like some of the more informal restaurants on Earth and no doubt will feature a big salad bar along with baskets of fruit. They can be located atop the spokes, just below the glass ceiling of the colony, to give patrons not only a panoramic view of their towns but also of the hang gliders flying from that level. The food will not be pretentious unless some fans of haute cuisine get together and decide to make it so. But with piped-in music and soft lights, the restaurants can have a pleasant atmosphere, so that prospective colonists back Earthside will hear about them and appreciate that the colony is not without amenities.

And for home entertainment, of course, people will have their stereos, their tape decks and TV's. There will quite likely be a brisk trade in copying tapes of current hits, since new records would prove difficult either to import (the freight rates again) or to manufacture within the colony. The colony will have its own radio and TV stations, which will originate some of their own programing, but which will mainly serve to rebroadcast transmissions from Earth. A single laser beam from Earth can easily carry the programing of all 83 TV channels, both UHF and VHF. One wonders how the colonists will react to "Star Trek" or "Space: 1999."

The communications systems will work in the other direction, too. It will be easy for people to make phone calls to friends back home, and everyone can have the use of the colony's WATS line (Wide Area Telephone Service) for free long-distance calls. This will be more than a casual entertainment. It will be one of the principal means of ensuring close ties between the colonists and the people of Earth. To achieve this is well worth the cost of sub-sidizing the phone system.

So, what will there be to do on a Saturday night? Quite a bit, pretty much whatever people want. As is true with so much else what there is to do will reflect the conditions at the colony and will be part of the development of a way of life which is distinctively that of the colony.

Those who return to Earth, for whatever reason, will carry the memories of good times with them. Anyone who has lived in those small closely knit communities will cherish memories of warm friendships, and more. Such a returnee may go swimming at a pool on Earth and look around. It won't be the same.

Chapter 12

THE SHELL OF THE TORUS

It was not a very large meteoroid as these things go, just a small chip from an old burned-out comet, a chunk thirty feet across. It had been knocked away at about the time some apelike creatures in Africa had begun to discover they could live on the ground, down from the trees—15, maybe 20 million years ago. Since then it had been orbiting in space, crossing and recrossing the orbit of Earth, yet never coming very close. Still, this could not go on forever. Someday it would certainly strike the earth, unless it hit Venus.

It was in that part of the orbit where it would begin to draw nearer Earth, feel its attraction a bit more strongly. This was nothing new, it had happened a hundred thousand times before. Space is big and the earth is small, a difficult target to hit with a rock moving randomly through space.

As the weeks progressed it moved closer. Had someone been aboard it, he would have seen the earth grow brighter and then begin to appear as a disk. Off sunward the moon would be distinguishable from the stars and in time it too would show a disk.

The rock was too small to see in even the largest telescopes. But could it have been seen and tracked, astronomers could have computed its orbit and determined where it was heading. Had they done so, they would have noted with some interest that this meteoroid would be passing relatively close by. Its path would be leading it between the earth and moon, almost midway between them, at a speed of three miles per second. The astronomers' interest would be only casual, however. Other rocks, large enough to see and track, had passed similarly close. There would be no significance to this new one. Within a year there would be other chunks of similar size on similar trajectories.

○   ○   ○

The navigation room in the space colony near the top of the central hub, adjacent to the dock-ing ports, controlled the small radar on the docking area which scanned surrounding space for in-coming craft. There was no need for either the navigation room or radar, but under the Space Stan-dards Act every manned spacecraft was required to have navigational facilities. For the purposes of this act, the Circuit Court of the District of Columbia had ruled that the space colony indeed was a manned spacecraft. There was no permanent officer on duty there and if there was, he probably would have been asleep. So no one saw the thin, light trace on the radar screen as the meteoroid headed in.

It struck the inner rim and passed through, leaving a neat hole. It was a fragile thing, easily broken, and the force of its impact was that of an explosion. It broke into several large chunks as it passed through the wall and these chunks fanned out as they ripped through the interior like frag-ments of a dumdum shell. When they reached the outer wall of the colony, they struck with the force of several tons of TNT. Where there had been Peachtree Park and the Apple Valley cluster of apartments there now was a hole 300 feet wide.

It took less than a minute for everything else to happen. The air in the colony began to pour through the gash at ten times the speed of a hurricane. For over a mile on either side of the hole, the rushing winds within the colony tore loose all buildings, carried off people and trees. The im-pact had been so sudden that there was no warning, no hope of warning. The shocks and blasts of the impact traveled round the colony at the speed of sound. But at the same speed came the destruc-tive hurricane–waves of rushing air swirling all into a tumbling, roaring, hurtling mass which gushed through the broken colony and into empty space.

On the side of the colony away from the impact, the winds were not so strong and the destruc-tion less severe. But in less than a minute there was too little air to breathe. No one had time to take action, none could save themselves. Some panicked, rushed outside, and were caught up in the winds. Many mercifully died in the wreckage of buildings. Where the houses did not fall and where people did not panic, there was time to head for a shelter but there were no shelters. Implacably, the air escaped, people gasped a bit, collapsed, their lips blue.

Several days later, the first ship of the Survey arrived. The colony was now surrounded by or-biting floating debris. The captain watched his radar and advanced slowly. Some of the debris drifted close enough to see clearly. There was the trunk of a tree– a palm, perhaps? It was hard to tell, its top had been sheared off. Again, there was what once might have been the roof of an apart-ment. But what made Captain Freitag shudder in later years, whenever he recalled it, was that other thing. It was a drifting form, limbs stiff, its face tight with horror and surprise–and with its torso cut nearly in two from a collision with some hunk of debris.

Someday, Hollywood will probably make a movie somewhat like this. Something like this may actually happen. The children of science who go forth to colonize space will find it a rich, fruitful medium in which to shape a new civilization. But nature does not yield its gains without risks and there will be hazards in space.

Probably the most frightening space disaster imaginable would be the collision with a swarm of large meteoroids, followed by explosive decompression and death for all aboard. This eventuality will probably only occur in future versions of such movies as *Earthquake*. The real meteoroid problem is finding minor leaks and repairing them in reasonable fashion.

The horrifying vision of a meteoritic disaster to a colony in many ways resembles the prospect of a nuclear disaster on Earth. Both involve arcane technologies with which few people have had direct experience. In both cases, people's convictions as to the smallness of the risks must rely heavily upon the testimony of experts, rather than upon the kind of common experience in which we adjust our driving to avoid a road mishap. Either disaster could strike with no warning, wiping out whole populations in a moment. The strange-looking domes of nuclear plants and the constant news reports of minor mishaps continually remind us of nuclear risks. In space, the occasional noticeable meteoroid impact, as well as astronomers' reports of small objects, will remind colonists of their own dangers.

But there is an important difference. We do not yet have so much or such thorough experience with nuclear plants to say for certain that their safety systems will always work. We can run test after test, pile safeguard upon safeguard. In the end, we largely come down to the fallible judgments of imperfect human beings. In the case of meteoroid dangers, we can see the objects of interest, accumulate statistics on impacts and sizes. We can go into space and measure the rates of impact by various types of meteoroid. What's more, we can construct a reasonably good understanding of how the probability of being struck depends on the size of the object.

In thinking of a rocky or gravel beach, there is some idea of the distribution of meteoroid sizes. There are a few large rocks and boulders big enough to be worth standing next to and have a picture taken. There are more smaller rocks and a great many smooth round stones or pebbles. More numerous are pieces of gravel. Far outnumbering all are the individual grains of sand.

So it is in space. The few large meteoroids, impressive enough to wind up in museums, fall at the rate of perhaps one a year. But on a clear night, you can go outside to look for shooting stars. You will probably spot at least a few, and each is caused by a meteoroid the size of a small sand grain.

We have data on meteoroid sizes and impact rates not only from observations made on Earth and in nearby space, but from instruments on the moon. These detect the shocks and small moonquakes which result when something hits the moon. The data from all sources tend to agree and give the following picture of meteoroid hazards to the colony:

Size	Mass	Rate of Occurrence	Worst Effects
1/4-inch pebble	1/100 oz.	Every 3 yrs.	Loss of window panel. Atmosphere escapes at 1% per hour.
2-inch rock	3-1/2 oz.	Every 7,000 yrs.	60% of atmosphere escapes in 10 hours.
Boulder	1 T	Every 250,000,000 yrs.	Severe damage.

The kind of catastrophe described at the beginning of the chapter probably would not occur before there was an entirely different catastrophe. The sun will expand into a red giant and engulf the earth—perhaps 5 billion years from now.

Anything heavier than a few hundred pounds would produce severe damage when striking the colony, about the same effect as setting off a fair-size bomb in the World Trade Center. However, even meteoroids as small as a pound would be rare and much less of a problem than airliners crashing into backyards on Earth.

The most common meteoritic effect would be a slow sandblasting of the windows and hull produced by micrometeorites. Over several centuries this would reduce the colony's bright metallic appearance, rendering it dull and gray and the windows would become less transparent, less clear. Once every few years the colony would probably be hit by something large enough to make people take notice. This would break a window, perhaps, or punch a hole in the hull.

On the scale of the colony, this would be similar to an auto puncture by a nail. A nail puncture usually results in a slow leak which you notice a few hours later when you stop for gas. The gas station attendant will cover the tire with soapy water and you will see the bubbles where air is escaping. It takes at least several hours for the tire to go flat, and similarly the leakdown rates in colonies will be quite slow. Daniel Villani of Princeton University has computed that with a hole as large as three feet wide, it would still take over ten hours to lose half the atmosphere of a colony.

Small leaks will be fixed much as you would repair a minor hole in a boat with a patch and caulking compound. For a larger hole, there probably will be something like a large metal tortoise shell which can be set over the hole to stop the leak. Workers in space suits then can weld an appropriate patch into place or replace a broken window pane.

There is a type of artificial meteoroid which may be dangerous to a colony. The mass-catcher described in chapter 6 would be propelled by rotary pellet launchers. The launchers resemble garden sprinklers which throw rocks—they eject small pellets at high velocities to give thrust. The pellets will not disappear but orbit through space as artificial meteoroids. They will not be sand grains, and will be just the right size for breaking windows or punching neat holes in the colony wall.

The pellets will not be ejected in any particular direction, but will be scattered and their hazard will be a matter for statistics. The astronomer George Wetherill, of the Carnegie Institute of Washington, has studied the danger from meteoroids with orbits similar to the pellets. He finds the pellets, like the meteoroid at the beginning of this chapter, will quickly move away from Earth's vicinity but stay between the orbits of Venus and Mars, crossing Earth's orbit repeatedly. After 10 to 100 million years, they will be swept up by Earth.

Earth has a billion times more area than a space colony. If we wish to have only one impact from a pellet every ten years, we can eject something like 10 quadrillion (10 followed by 15 zeros) pellets. This is a million times the mass of pellets to be ejected in transporting material for the colony.

The hazard from meteoroids will be very small and the environmental effects of the rotary pellet launcher will be entirely lost in the natural effects due to meteoroids. There is little danger with large particles of space matter, those big enough to hold in the hand. Unfortunately the same is not true of the smallest particles of space matter, the cosmic rays.

Cosmic rays are charged particles which stream outward from the sun and inward from the galaxy. They are predominantly the nuclei of atoms of hydrogen, moving at close to the speed of light, but it is likely that every element of the periodic table is included in their nature.

When the sun is quiet, it emits a solar wind of protons and electrons which stream outward at speeds of 300 miles per second. The particles do not penetrate and pose no threat to the colony. However, every eleven years the sun enters an active phase and can produce solar flares. In these sporadic violent eruptions, the sun emits blasts of high-energy protons which can deliver dangerous doses of radiation. In the flares the particle energies usually are below a billion electron volts, but the worst radiation conditions occur during eruptions similar to the great flare of February 23, 1956. The proton energies then can range up to several billion electron volts, which gives approximately the radiation environment inside a research-type nuclear accelerator. An unprotected human being would need less than an hour to receive many times the fatal dose of radiation.

Cosmic-ray flares occur once every several years, and flares as large as the 1956 event occur every few decades. Because many of their protons travel at nearly the speed of light, there are only a few minutes between detection of the flare by telescopes and the arrival of the worst of the cosmic rays. People not in a sheltered place will have very little time to get to one. Once a flare begins, its most dangerous time lasts for somewhat less than a day, as streams of energetic particles course in all directions.

Cosmic rays are dangerous because, being electrically charged, they break chemical bonds when they pass through tissue. This can create damage to cells which will lead to cancer or damage chromosomes to produce genetic mutations. The damaging power of a cosmic ray is related to its "ionizing power," which measures how many chemical bonds are broken. Ionizing power does not increase with energy as might be expected. The most energetic particles, moving close to the speed of light, pass swiftly through the body and do relatively little damage. It is at lower energies, speeds less than the speed of light, that the particles have more time to break bonds and do much greater damage.

If we calculate the annual radiation dose which the general flow of cosmic rays would give to an unshielded person (aside from the occasional solar flare), the dose is about ten rem per year. The rem is a unit of radiation exposure used in the nuclear industry. This is not a high dose. It is only twice the dose for radiation workers recommended by the Atomic Energy Commission for adults working in industries where exposure to radiation is likely to occur. There is evidence that exposures to steady radiation levels up to fifty rem per year will not result in detectable damage. If this total dose were the only problem, then the solution would be a simple matter of providing light shielding over the colony or construction areas, along with heavily shielded shelters to retreat to during the infrequent solar flares. But the problem is more complex.

Included in the cosmic rays are substantial numbers of iron nuclei totally stripped of electrons. When a fully ionized iron nucleus is traveling at less than half the speed of light, its ionizing power is several thousand times that of an ordinary proton. Passage through the body of a single iron nucleus destroys an entire column of cells along its path. The total amount of energy which the particle dumps into the body is small, but it is highly concentrated. This radiation can not only increase the risks of cancer, it can provide pathways of damaged or dead cells along which the cancer can spread and grow.

The iron nuclei do even more. They destroy nerve cells in the brain and spinal column which cannot reproduce themselves. Once the cells are dead, they can never be replaced. Studies of Apollo astronauts indicate that on their two-week lunar voyages, they may have lost as much as one ten-thousandth of their nonreplaceable neurons. Under such conditions after several years in space, the loss could reach several percent. For children, especially, the effects could be devastating.

It is also important to be aware of the phenomenon of secondary particle production. When high-energy particles collide with matter, as in a radiation shield, they give off a spray of particles. These in turn may produce more particles. In the presence of very energetic particles, therefore, a little shielding may cause an even larger radiation dose than if there were no shielding. There also is the possibility that a little shielding will slow down fast iron nuclei, making them more damaging to tissue.

How can we protect the colony against radiation? The first thing is to set a standard. Actually, the U.S. government sets two standards now. For radiation workers—that is, for adults over eighteen—the standard is five rem per year. For the general population, which includes children and pregnant women, the limit is one-half rem per year. This limit is about the level of radiation which occurs naturally in cities of high elevation like Denver. In addition, this standard provides that the colonists will receive no more iron nuclei over thirty years than Apollo astronauts received in two weeks.

To reach these limits the same means of protection which the earth provides can be used. The earth has a magnetic field which deflects many particles and an atmosphere thick enough to shield us from those that get past the magnetism. Both will serve for protecting the people of a space colony.

For the construction shack and the colony's external work areas, the solution might be to enclose the construction spheres in a magnetic field. When a charged particle passes through a magnetic field its path curves, and an appropriate arrangement of magnetic fields can form a region where particles cannot enter. Not all particles will be deflected this way, however. High-energy particles will curve only slightly and penetrate the shielded region.

The construction workers can receive adequate protection by designing the shield to deflect particles with energies below half a billion electron volts. The particles deflected would include those from most solar flares as well as the worst of the cosmic rays (including the iron nuclei). The shield can be kept light enough for construction spheres of reasonable size not to produce secondary particles to any major degree.

On rare occasions solar flares would burst forth with an intensity that would make a shield inadequate. For those times the construction areas should be equipped with flare shelters made of enclosed modules with thick walls. At first these walls would be filled with a foot of lead and carbon to stop radiation. Later, the valuable carbon could be removed for use in the colony, with lunar rock substituted. Gerry Driggers, who has done much of the work on construction shack studies, has described the Strangelovian life of people in these flare shelters:

> Each [shelter] would be equipped with three hundred comfortable reclining seats, reading material, movies, and some food and drink. . . . Stay times could be made practically indefinite with brief excursions for food and water. . . . One square meter has been allocated per man with individual reclining seats provided, each equipped with a foldaway desk. Entertainment would consist of reading matter, movies, and various games which can be played in a small area. The shelter would naturally double as the primary recreation area for off-duty crew members.*

However, Dr. Strangelove, of the Stanley Kubrick movie, proposed to have people live under such conditions for ninety-three years, and Dr. Driggers anticipates that "the infrequent use of the shelter will probably be for stays of three to eight hours."

For the colony itself, all this is insufficient. The exposure limit, one-half rem per year, cannot be attained simply by excluding the low-energy particles and by keeping flare shelters handy just in case. Instead, it is necessary to keep out all particles with energies up to 10 billion electron volts.

There are no more than about half a dozen laboratory accelerators which can produce hard radiation with such energies. The hard-radiation areas of the accelerators are buried

---

*From a paper entitled "A Baseline L₅ Construction Station," presented at the 1975 Princeton Conference.

under many feet of rock and soil to protect passers-by. To protect the colonists against such particles will be difficult.

It is not possible to use magnetic shielding. The coils of superconducting wire and cable, the cooling apparatus, most of all the structural steel to withstand the magnetic forces—all these together would weigh several million tons. A magnetic shield for the Stanford torus would not be a simple set of coils wrapped around an enclosed space. A shield like this would, indeed, *be* the colony.

Though magnetic shields will be too heavy, the same may not be true of electric shields. If the colony were to be charged electrically to 10 billion volts, it would repel cosmic rays very effectively. Ten billion volts would repel all particles with energies below the desired cutoff—10 billion electron volts.

While virtually all cosmic rays would avoid the Stanford torus as if it were a leper colony, every electron for hundreds of miles around would feel a strong attraction. They would rush toward the colony with the same high energy, 10 billion electron volts. When they got there, they would do two things, both of them bad. The first would be to produce strong and damaging X rays, which occur when energetic electrons strike a metal surface. The second would be to neutralize the electric charge. The colonists would be no better off than if they had no shield at all. They would actually be worse off, since they would all be dead from the X rays. So if electric shielding is to work, it is necessary to be a bit more clever.

Eric Hannah, Gerry O'Neill's associate at the 1975 Summer Study, thought of a possible way out. Unlike cosmic rays, electrons are very light and are easily deflected by small magnetic fields. Suppose, he said, we were to build the hub of the colony around a hollow core wound with coils carrying electric current (physicists call such an arrangement a solenoid and often use solenoids to produce magnetic fields); it would produce a field adequate to keep the electrons from reaching the colony. The electrons would all be funneled into the interior of the hollow core where the magnetic field (hopefully!) would keep them from ever reaching the walls. With this arrangement, an electron gun (having roughly the power of the two-mile-long Stanford linear accelerator) could pump out enough electrons to maintain the charge on the colony.

This is the kind of speculative technology that may become attractive once we have learned as much about controlling electrons in free space as we know about controlling electrons in wires. But no one will risk a colony and its people on the off-chance that the electrons in the solenoid will never, never find a way to reach that colony hull to which they are attracted so strongly.

There is an easier way to shield the colony and it is beautifully simple. Just shovel 6 feet of lunar soil around the whole of it. Let the colony hull be six feet under such a covering and all radiation problems will vanish.

The six feet of covering will absorb all cosmic rays as well as most secondary particles. A few of the most energetic particles will produce so many secondaries that some will reach

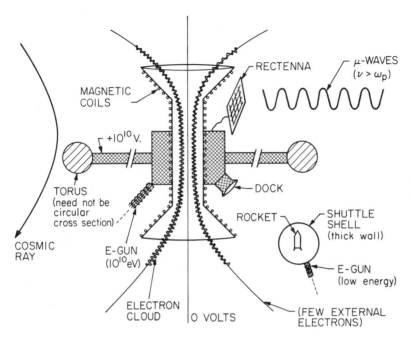

Plasma radiation shield proposed by Eric Hannah. Cosmic rays are repelled from the colony, which is charged to 10 billion volts with an electron gun. The magnetic coils channel electrons from space into the central region, and prevent them from reaching the walls, which would neutralize the charge. (Courtesy Eric Hannah)

the colony, but there will be no problem with solar flares, even the biggest, and no problem with the iron nuclei in cosmic rays. Babies and mothers will live amid radiation levels no higher than those in Denver or in some of the cities in northern Arizona.

But six feet of soil over all 440 exterior acres of the Stanford torus is a rather large pile. It is 10 million tons, every pound of which will have to be shot from the moon by the mass-driver, then delivered to the colony by the mass-catcher. It is this thick shell, this radiation shield, which will certainly be the most massive single item to construct.

It was late in the 1975 Summer Study when we realized this. For a day or two that August, there was serious doubt that we would be able to accomplish our mandate: "Design a system for the colonization of space."

But it wasn't long before this shield began to look feasible, and soon it began to look almost inevitable. For much of the 10 million tons will simply be the slag or waste left over after the metals have been taken from the lunar soil. On Earth, the processing of low-grade ore inevitably produces huge piles of useless tailings, dumped into a lake if possible, a hole in the ground if necessary. In space, the same tailings will not be waste. Some will serve for making glass, some perhaps as a source for oxygen. But most of this material, resembling good beach sand, will be used to fill the radiation shield.

This shield will be a hollow shell of aluminum, fitted around the Stanford torus in the

Space colony with surrounding radiation shield. The artist (Chizanskos) has used license in depicting the shield as built of large rocks, like a breakwater; actually, it would be a loose fill of lunar soil or rock within a shell surrounding the torus. (Courtesy NASA)

way that a bicycle tire fits around an inflated inner tube. This shell will be filled with the sandy slag or tailings, blown in much as wheat is blown into a grain elevator in the Midwest. When a mass-catcher arrives, a part of the shell will provide a convenient spot to store its hundreds of thousands of tons of lunar ores. The construction workers, building power satellites or new colonies, will draw on these stored reserves. They will process them, take from them the needed metals and glass and oxygen and return the tailings to the shell.

The shell will not be quite the same as a bicycle tire in relation to its inner tube, since it will not rotate. The Stanford torus—the inner tube—will rotate once a minute. But if the shell rotated it would need to be built much more strongly than is reasonable. So it will not rotate, but will enclose the colony hull as the hull rotates past at over 200 miles per hour. There will be a clearance of several feet to keep the hull and shell from touching or scraping. The shell will be tied by cables to the central hub to ensure that no such scraping can occur and the cable lengths will be maintained tight like bicycle spokes. Inevitably, there will be concern that the hull and shell may scrape despite these precautions. The solution will be simple: add more clearance. An extra 4 feet of clearance will only increase the shell mass from 10,000,000 tons to 10,100,000.

How can light get through to the colony under such a shield? The colony mirrors can reflect sunlight down into the Stanford torus. Sunlight cannot be provided by making the shield only a partial shell, leaving a gap. Through the gap would come rays much more penetrating than sunlight. Sunlight, however, reflects off a mirror, but cosmic rays do not. So where sunlight is to come through, we build the shell, not as a solid covering, but as a collection of overlapping V-shaped sections resembling lengths of angle iron. The flanges of these will be the standard six feet thick, with their surfaces covered with mirrors. The sunlight will

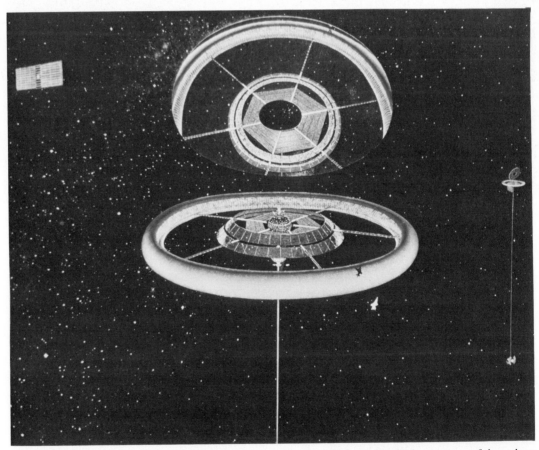

Overall aspect of the space colony. The top structure is a large mirror, in which an image of the colony is seen. A second colony, at the right, shows how the Stanford torus is connected to the extraction facility. A power satellite is at the upper left. (Donald E. Davis painting courtesy NASA)

Installation of chevron shielding in the colony. The chevron shields are the structures resembling lengths of angle iron. (Donald E. Davis painting courtesy NASA)

reflect from mirror to mirror and on inside, but the cosmic rays will be absorbed in the plates of the V's. The arrangement of these V's, stacked as they are one above the next, resembles the stripes of a sergeant's chevrons. It is this "chevron shield" which solves the problem of letting in the light.

When we have protected the colony, we find we have actually recreated the protection we enjoy on Earth. The atmosphere above us gives us the protection we would receive from a shield of lunar soil ten feet thick. The colony's protection will be a bit less, but will follow the same principle.

Inside this protection the colonists can settle down to build their lives. They can establish schools, organize clubs and churches, in short, do nearly anything they would do back on Earth. Including going to the movies to see a film about a meteoroid hitting a space colony.

Chapter 13

UNIVERSITY OF SPACE

Tucked away in the spokes of the colony, along with the cattle feedlots and the air-conditioning systems, will be the colony's principal center for scientific research. People living in their apartments will have for neighbors mostly construction workers or agricultural workers; but here and there will be an infrared astronomer or an expert in communication with extraterrestrials—or an anthropologist, living as unobtrusively as he can while studying the natives. The colony will be more than a place to build powersats, more than a settlement. It will be a place to do good science, to produce the excellent papers whose authors will give for their affiliation "The University of Space."

How will this university be organized? As in the great Earthside research centers, there will be permanent and senior faculty as well as a number of people on temporary appointments. The permanent positions will belong to those scientists who will have been involved from the start in establishing the university. However, if they are like senior scientists in the Earth's research centers, they will spend little time in the laboratories or at the telescopes. Instead, they will be attending meetings with senior colony administrators or going to Earth to testify before Congressional committees or meeting with wealthy donors to seek funds.

The real work of the University will be in the hands of the junior faculty or the researchers there on temporary fellowships. These will be eagerly sought by Earth's best graduate students and young researchers. Some of these people will so distinguish themselves that they will be invited to stay, to join the growing faculty permanently. But when they first arrive, like junior researchers everywhere, they will find themselves sharing cluttered offices and labs.

Much of this work will be quite specialized, of interest mainly to the scientists who are doing it, and to their close associates. For instance, there will be the zero-g biology lab. There small animals will be used for experiments while scientists earnestly study their blood counts and bone calcium levels, or allow them to breed entire generations that will never feel gravity. Then their keepers will write papers for *Science* with titles like "Sexual Response in Guinea Pigs: Effects of Zero G."

Zero g will also provide the opportunity for a lot of work in the general area of materials and chemical processes, some of which may lead to new industrial ventures which will return a tidy profit. For instance, very large crystals can be grown of the type needed by the electronics industry. You can purify molten materials by a technique known as "zone refining," which works best without gravity. Many people have noted the possibility of producing perfectly round ball bearings by allowing molten metal droplets to form spheres under their surface tension, then to solidify. There is a process called "electrophoresis" in which you can separate materials in solution by having their molecules migrate under an electric field, applied across the container. It is useful for purifying vaccines and other pharmaceuticals, and it works best in zero g. Finally there is the ability to mix materials in the liquid state which would ordinarily separate, like oil and water. Foam steel (like foam rubber), alloys such as iron-lead with unusual electronic properties, composite materials—all can be made in zero g.

These are some of the immediate possibilities. In the long run, the work of the zero-g lab may produce advances in science of the highest importance. Our present highly limited experiences with zero g may fail even to hint at the discoveries to be made. We are so pervaded by gravity, we find it so hard to escape its effects, that in thinking today of the prospects of zero g, we can only be like fish living in the sea, wondering at the possible uses of fire.

We may gain a hint as to the future importance of zero g by comparing our situation now to what happened three centuries ago, when it became possible to remove another pervasive force—air pressure. The ability to make a vacuum led to the steam engine and to other engines and to many of the inventions and developments of the nineteenth century.

But the ability to make a vacuum did even more. In the last years of the last century and the early years of this century, physicists were astounded by a variety of new phenomena. There were electrons, ions, X rays, atomic structure, radioactivity and much more. And what lay behind many of these discoveries? It was the humble vacuum pump, finally improved to the point where it could produce vacuums sufficiently good to allow electrons and ions to show their true behavior. Not only did these discoveries spark the development of much of modern physics; they also led, early on, to the vacuum tube, radio, and the whole of modern electronics.

If someone had asked Torricelli, the inventor of the vacuum pump in 1650, what his invention was good for and if he had answered, "It opens the age of engines, electronics, and nuclear physics,'' no one would have understood or believed him. We cannot now imagine what zero g will mean to the people or the scientist of the year 2300—or even 2100.

In zero g materials in the liquid state become objects in their own right. They can be studied and heated or treated without a container, or held in place with small forces. This means it will be easy to study highly corrosive materials or materials of very high melting point. They can be produced and studied in very pure states, since there will be no danger of their dissolving any material from a container.

On Earth, liquids can almost never be studied entirely apart from the containers which hold them. Thus, physicists have been unable to develop a satisfactory understanding of the liquid state to match their knowledge of the solid state. In zero g, where the most subtle effects of the liquid state can be observed, all this will change. Solid-state physics has given us much of our present-day electronics. What may we gain from liquid-state physics?

And the unprecedented control over materials, their purity, their mixture with each other, their composition, may well herald an era when we will routinely create metal-glass mixtures or plastic-ceramic alloys with virtually any properties desired. There is an immense range of properties available from the products of today's plastics industry. This may be only a taste of the range of properties available from the zero-g industries which, like the electronics firms near MIT, may grow up around the colony's laboratories.

We may obtain superconductivity at room temperature. A superconductive material has zero electrical resistance so that it can carry immense flows of current, or give rise to huge magnetic fields. But at present this highly prized property can only be achieved at very low temperatures, such as the temperature of liquid helium. The development of a room-temperature superconductor would completely transform transportation and could make the wheel nearly obsolete. Cars and trucks, instead of being supported by tires, would have small, powerful superconducting magnets. Their ''roadways'' or guideways would be of aluminum and, when traveling faster than ten or twenty miles per hour, the magnets would support them precisely as the buckets on the mass-driver will be supported.

Beyond this, our terrestrial ideas of the possible practical size of processing equipment and facilities may need adjustment. The disappearance of dead weight in machinery will allow us to build tools of truly extraterrestrial dimensions, perhaps capable of using even cosmic phenomena as production means.

But zero g is only one of the phenomena which will lure scientists to the colony. Another one is the entire surrounding sky, visible with unprecedented clarity. It is certain that the colony will quickly become a leading center for astronomy.

Since the 1950s, the development of new instruments and techniques has wrought a revolution in astronomy. The discovery of quasars, pulsars, black holes, and the background radiation from the birth of the universe, has pushed astronomy again to the position it held in the time of Newton: at the forefront of the physical sciences, and among the studies promising the most important new insights into nature. The past twenty years have seen major new additions to the facilities at existing observatories, such as Mount Palomar. Whole new observatories have been built, as at Kitt Peak and Cerro Tololo. Radio astronomers have built their immense receivers, shadowing acres of ground in California and West Virginia, Australia and Puerto Rico.

The history of astronomy is a history of efforts to overcome three types of limits upon the ability to observe objects in space. The first is the limit set by the dimness of these objects. To see fainter and fainter objects, astronomers have traditionally built larger and larger telescopes to gather more light. But on Earth gravity limits the size of telescopes. Much larger telescopes can be built for use in space, where gravity presents no problem.

The second limit is a limit on the sharpness or detail of an image, what astronomers call resolution. It is of little value to form a bright image of a distant object if that image proves to be blurry, and the atmosphere blurs images. The importance of resolution is illustrated by comparing the bright but blurred images of Mars taken with Earthbound telescopes with the far clearer photos taken at close quarters by spacecraft.

The third limit involves the range of wavelengths which can be studied. Until the start of the revolution in astronomy, for all its big telescopes and sensitive photography, astronomy was still as keyed to the human eye as it had been in the time of the Babylonians. The first big advance, outside the range of wavelengths accessible to the eye, came when Ryle in England and Mills in Australia began to build radio telescopes. Today, radio astronomy is fully on a par with optical astronomy in terms of the sophistication of its instruments and the number of talented people it has attracted. More recently, advances in infrared sensors led Gerry Neugebauer of Caltech to set up the first major infrared observatory; and satellites and spacecraft have begun to give astronomers tantalizing glimpses of the X-ray sky, the gamma-ray sky, and the far ultraviolet.

However, at radio wavelengths, man-made interference can be a serious problem. Radio astronomers have a saying that when they see anything interesting ninety-nine times out of a hundred it turns out to be someone's electric shaver or auto ignition. Also, Earth's atmosphere will continue to block out or absorb many of the most interesting wavelengths.

All these problems can be overcome in space. What is needed is a permanently staffed observatory in space, supported by a community which will fully recognize its value. The space colony will be such a community, and the University of Space will have such an observatory.

At the colony the same control over material properties which will yield new alloys also will permit the casting of optical blanks for telescope mirrors of unprecedented size. These may be used with image tubes, which amplify faint light so that an exposure which formerly would have taken hours now can be done in a few minutes. The resulting telescopes will not only excel in viewing faint objects but also provide fantastic resolutions. In seeking larger telescopes one useful technique will be the "rubber mirror." This is an array of small mirrors, adjustable in position to give the performance of a single large one.

Orbiting telescope, of a type suitable for use in a major colony observatory. (Courtesy Boeing Aerospace Co.)

Radio interferometry at extreme resolution will be put on a permanent routine basis. Standard radio telescopes will be simple light affairs, perhaps resembling giant umbrellas rather than requiring the massive trussworks and supports of their Earth-based cousins. There may be many such instruments, widely separated in the space near the colony, yet always within sight of each other and of the colony—a situation which makes it much easier to do the interferometry. They can be shielded with wire mesh against stray signals from Earth or the moon. Other radio telescopes will be built, miles in extent, to detect the faintest signals from the most distant sources.

The X-ray and ultraviolet astronomers will especially benefit. No longer will they need to wait patiently for the few opportunities each decade. Instead, they will simply set up a workshop in an odd laboratory of the university and proceed to build instruments to their desires. As with the optical shops at Pasadena and Palomar, such a lab will give flexibility and convenience to their work.

Few branches of astronomy will fail to benefit from these advances. Much of the Milky Way galaxy lies hidden behind dark nebulae such as the Coal Sack; this includes the interesting regions near the center. Yet radio and infrared waves readily penetrate these clouds. Thus we may become as familiar with the Milky Way as we are with the solar system. We will penetrate with high resolution into the heart of the Orion nebula, where stars are being formed, to learn the details of their origin.

The most distant galaxies and quasars may be mapped and studied as if they were a thousand times closer. In the vast drowned depths of space and time, newly accessible to our instruments, we will see directly the effects of the curvature of space. But when we look outward 20 billion light years in space, we are also looking at events which occurred 20 billion years ago in time; and so we may look backward to the very morning of the day of creation.

Closer to home, there will be a great flowering of the planetary sciences. The high resolution and breadth of observable spectrum will permit spaceborne telescopes to do much of the work now assigned to spacecraft in studying the planets and small bodies of the solar system. Even in an era of space colonization, planetary missions will continue not only to be costly but also to take years to reach their targets. Besides, there are such a large number of interesting objects that it will take a long time to visit them all. Some, like comets and the most distant planets, may not be visited for many years.

With the best spaceborne telescopes it will be a matter of routine to see objects two or three miles across on Mars or features of that size in the atmosphere of Venus. Jupiter and its satellites will be studied at a resolution of a few tens of miles, while hundreds of asteroids and comets may be mapped at least as accurately as were the Martian satellites, Phobos and Deimos, through the photography of Mariner 9. The roiling storms of Jupiter will be seen as clearly, and as routinely, as the hurricanes of Earth. Even distant Pluto and the satellites of Uranus and Neptune will be mapped with better precision than was Mars prior to that same Mariner 9.

A "cup-and-saucer" radio telescope such as can be built adjacent to a space colony. The "cup" is the radio reflector; the "saucer" is a shield against radio interference from Earth. The free-floating objects are radio feeds. An advanced space shuttle indicates scale. (Courtesy NASA)

The "cup" and "saucer" in a different orientation. (Courtesy NASA)

Routine maintenance on a thrustor unit of the "saucer." The "cup" is in the background, to the right of the space shuttle. (Courtesy NASA)

A particularly important activity will be the detailed study of the sun. The solar astronomers will be concerned with developing a science of solar meteorology, to predict the solar storms which will put spaceships and unshielded quarters at hazard. (Indeed, this activity will be one of the major reasons why the colony will support its astronomers.) But beyond this, there will be efforts to understand and predict slow changes in the sun's behavior which can influence the climate of Earth. A single well-founded prediction—"The mean global temperature in the next ten years will be 0.3 degree lower than in the last ten years"—will have such implications for Earth's agriculture as to be worth the cost of 100 major space observatories.

But all these activities will be extensions of work already under way in astronomy. The most significant findings may come from studies such as we cannot now undertake but which will become possible at the colony's observatory.

The detection of extrasolar planets may be among the easiest prizes. Even with the highest resolutions, it will be difficult to detect a planet as large as Jupiter against the glare of its nearby star—much less to detect a planet like Earth. But there is a method which will work. Stars do not stay accurately fixed at the same point of the sky, but move slowly in relation to more distant stars with what is called their "proper motion." If a star has an unseen companion in orbit about it, the star will be seen to move from side to side along its track. Its motion will exhibit wiggles or, as astronomers call them, "perturbations." From a study of these perturbations, it is possible to determine the mass and orbit of the companion.

The difficulty is that detecting extrasolar planets is a task at the limit of today's capabilities in astronomy. Today's, yes—but not tomorrow's. In the colony, all stars out to several dozen light-years will be routinely tracked and studied for evidence of perturbations.

Two galaxies drifting past one another, as seen in the sky of an Earthlike world. Relative to their size, galaxies tend to be almost frighteningly close to one another. (Painting courtesy Don Dixon)

Other methods as well will be used, like sending a small guided sphere thousands of miles out, to block the light of a star. In such an artificial eclipse the dim light of a planet will be more easily seen.

The single most exciting quest, however, will be for intelligent life in the universe. This search, even today, has attracted enough interest to be granted a four-letter acronym for the use of the bureaucracy—SETI, Search for Extra Terrestrial Intelligence. It has an in-group meaning, too, for it brings to mind the star Tau Ceti, twelve light-years away and very much like the sun, hence possibly the center of a planetary system containing life. NASA has seriously considered making SETI one of its major research themes for the years to come. And it appears likely that long before the first space colony is up there, there will be serious ongoing work devoted to SETI. The colony then will serve to extend this work.

SETI people nowadays are most interested in the "Cyclops" system, proposed by Bernard Oliver of Hewlett-Packard, as a means to carry forward the search. Most of the attention given to Cyclops has focused on Oliver's proposal to build an immense array of radio telescopes, miles in extent, for the search. (This was the emphasis in Arthur Clarke's *Im-*

perial Earth.) But the heart of Cyclops is nothing so dramatic. It is a means for rapidly and accurately scanning an entire broad region of the radio spectrum with high sensitivity to detect a weak drifting signal amid the cosmic noise. This technique involves an optical spectrum analyzer, which simultaneously receives signals on up to a billion distinct wavelengths and records the signals on photographic film. By searching through narrow bands at neighboring frequencies, Cyclops will detect a signal a billion times weaker than the surrounding noise in which that signal is buried.

But at what frequencies shall we search? Here too, Oliver has a recommendation. At the lowest frequencies, in the hundreds of megahertz, there is strong interference due to radio noise from the galaxy itself. At frequencies of a few gigahertz (billions of cycles per second), the background noise from the Big Bang, from the universe itself, begins to pick up. At slightly higher frequencies radio signals are absorbed by water vapor in the atmosphere. So the best part of the spectrum turns out to be from about 1 to 2 gigahertz.

It happens that right in this region there are two cosmic radio spectral lines: at 1.42 gigahertz the strong line due to interstellar hydrogen and a bit higher, at 1.66 gigahertz, another line due to the hydroxyl ion. These lines occur in a region of the spectrum where the noise is lowest, where there is a "hole" in the noise spectrum. Also hydrogen and hydroxyl are produced by chemical breakup of the molecule of water. Oliver regards the band between the spectral lines of the dissociation products of water as the place where water-based life may search for its kind. "Where shall we find other intelligent species? Why, at the age-old meeting place of all species—the water hole."

What would we learn from SETI? There may be a community of communicating civilizations of which we would be merely the youngest member. This community may have built up a vast body of knowledge over the eons, what Oliver calls the galactic heritage:

> *The galactic heritage could include a large body of science that we have yet to discover. It would include such things as pictures of the Galaxy taken several billion years ago; it would include the natural histories of all the myriads of life forms that must exist in the planets of their member races. We could see the unimaginable diverse kinds of life that evolution has produced in other worlds and learn their biochemistries, their varieties of sense organs, and their psychologies. Culturally, we might learn new art forms and aesthetic endeavors.*
>
> *But more significant will be the societal benefits. We will be in touch with races that have achieved longevity. The galactic community would already have distilled out of its member cultures the political systems, the social forms, and the morality most conducive to survival, not for just a few generations, but for billions of years. We might learn how other races solved their pollution problems, their ecological problems, and how they have shouldered the responsibility for genetic evolution in a compassionate society.*
>
> *The best outcome of Cyclops is something we cannot yet conceive. I can only suggest that childhood's end may await us at the water hole.**

---

*From the December 1974–January 1975 issue of *Engineering and Science,* the bimonthly house organ of the California Institute of Technology. **175**

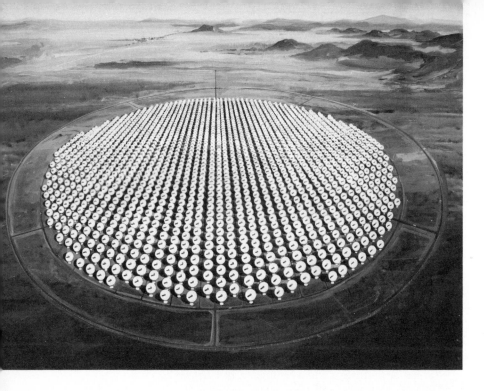

Overall view of the Cyclops system: an array of large radio telescopes, filling a five-mile circle. (Courtesy NASA)

View of Cyclops from just above the radio telescopes. (Courtesy NASA)

View of Cyclops from ground level. Each radio telescope is 300 feet in diameter. (Courtesy NASA)

Simulated reception of an intelligence-bearing signal with Cyclops. Each horizontal line represents a frequency observed. The passage of time is indicated by the sweeping from left to right, at each frequency. White streaks are noise in the receiver. The signal is so weak that if it were observed at only one frequency, it would not be distinguished from the noise; yet at adjacent frequencies, there is a clear pattern. The slope of the signal to the right results from delays, at adjacent frequencies, due to the effects of the interstellar medium; the angle of the slope indicates the distance to the source of the signal. (Courtesy Martin Ewing, California Institute of Technology)

Chapter 14

THE NEXT MILLION YEARS

Sir George Darwin, son of the famous Charles, predicted that our present era would be seen as a golden age compared to the vistas of famine and poverty which would follow in times to come, as Earth's teeming billions fought over its waning resources. Darwin echoed the classic arguments of all neo-Malthusians envisioning mankind's fate in terms of human procreation in a finite world. However, he did suggest that the situation would at least be improved, perhaps even solved, granted two prerequisites: population control and an inexhaustible energy source. Somewhat similar views had been expressed by H. G. Wells in his 1914 book *The World Set Free,* in which he predicted the development of the atomic bomb.

We will take quite a different view of the long-term future. Barring a catastrophic epidemic of human stupidity, the decades ahead are likely to see the foundations solidly laid for a world without large-scale poverty or hopelessness, a world of opportunity, rising living standards and widely shared middle-class levels of affluence. Such a world will endure into the indefinite future. It will not be without problems or difficulties and there will continue to be challenges aplenty; it will not be a utopian dream of equality and selflessness. The world of the next century will be one in which most people live at least as well as in today's America or Europe.

Such a world will not be achieved easily, for it will be necessary to solve such difficult problems as the population explosion. It is no mere neo-Malthusianism to be concerned about the rapid growth of world population, for it can seriously delay the improvement of living standards over much of the world, and it is necessary to be aware of its actual nature. Mere projections of current growth rates will not do, nor will predictions that within a few centuries the world will resemble Yankee Stadium during the World Series. These fancies come from people who are not prepared to consider what enters into that intimately personal matter, a couple's decision to have a baby, but who do know how to use the log-log scales on a slide rule.

As is true with so much of this planet, there is no "world" population. It is much more useful to speak of the developed nations and their populations, as distinct from the underdeveloped ones. The developed countries include the United States, Canada, and most of Europe as well as Japan, Israel, and the Soviet Union. There are 31 in all, according to the population analyses of Charles F. Westoff of the Princeton University Office of Population Research. Together they account for 27 percent of the world population.

It appears that a simple statement can be made which characterizes the problem of overpopulation or too-rapid population growth in the developed countries: It is not a problem.

In the United States in recent years there has been a sharp falloff in the rate of childbearing. The rest of the developed world, less well known, has generally had a similar decline. To achieve zero population growth it is necessary that, on the average, women have 2.1 births in their childbearing years; each generation then will replace the next. In 1973 in the 31 developed nations, the corresponding figure was 2.2. In 20 of the 31, including the United States, Japan, and much of Europe, the rate is close to or below 2.1 and population trends are in the same direction in most of the remaining 11 countries. Only Spain, Portugal, Ireland, and Israel still have rates much above 2.5.

The currently low rates do not mean that zero population growth is around the corner. Populations will continue to grow for a few more decades because most societies still have proportionately more younger people in their childbearing years. However, there is little prospect of returning to higher rates of birth. In a number of developed countries, recently married women have been polled to find how many children they wanted or expected to have. The answers ranged from a high of 2.2 in the 1972 survey in the United States to a low of 1.8 in England.

The long-term prospect for the developed world seems to be similar to what France has experienced for the past two centuries. At the time of the French Revolution, its population of 25 million made France the most populous state in Europe. It easily stood off invasion by a coalition of powers determined to overthrow the Republic, and under Napoleon proceeded to a career of conquest. But in all the nineteenth century, its population grew by only 12 million. Germany quickly surpassed France, with unfortunate results for the peace of Europe. Today, at 52 million, France has maintained an average growth rate of only 0.4 per cent per year since 1789.

In the underdeveloped countries, the situation is very different. The population is truly exploding. In Latin America, in the early 1970s, the population grew at 2.7 percent per year, so it would double in 26 years. In Africa the rate was 2.6 percent, and underdeveloped Asia, excluding China, grew at 2.4 percent. China, which may be beginning to control its population growth, may have grown at 1.7 percent.

These extraordinary rates apply to nearly three-quarters of the human population. They do not result from recent large increases in the birth rate, quite the contrary. Birth rates in the underdeveloped world have been steady in recent years or, in some countries, begun to decline. What has happened is the introduction of modern medical techniques which have controlled many diseases, reduced infant mortality, and increased life expectancy. The falling death rates, unaccompanied by comparable reductions in the birth rate, have produced the resulting rapid growth.

This situation will not continue indefinitely. It results from the lag between the drop in death rates, which can be achieved through technology, and a drop in birth rates, which requires profound social changes. Longer life is recognized as desirable in all societies, but high fertility also has usually been regarded as desirable. It is the change in the latter viewpoint which takes time to accomplish and which produces the lag. However, there are several influences which in time can change people's views of the advantages of children and reduce the birth rate to a more manageable level.

A powerful incentive in reducing conceptions is people's unwillingness to accept a lowering of their accustomed standard of living. In nations like India this may be the most important influence. In many underdeveloped nations, however, programs of development have begun to raise the standard of living. In these societies people begin to find it possible to acquire bicycles or other consumer goods, to provide a better life to children already born, or to raise their status in society. Their chances of doing these things are often powerfully increased by having fewer children.

Overall, there will be continued and growing influences promoting reduction in the birthrate, while influences of old ways of living will tend to keep it high. Their power cannot be discounted but with the rapid pace of change in today's world, the universal desire for development and for better living standards, it can scarcely be doubted that the underdeveloped world also will achieve population control. But before the current population boom runs its course, the population of the underdeveloped world will quite likely double and then double again, during the next century or so.*

This raises the question of feeding the hungry masses. According to Roger Revelle, director of Harvard's Center for Population Studies, the earth's arable land can probably provide food for 40 to 50 billion people. However, this would happen only if the land were tilled using the advanced methods of Western agriculture. It is a prime goal of many develop-

---

*In November 1976, new population growth rates and projections were made available. Lester Brown of the Worldwatch Institute stated that the overall world population growth rate, 1.9 percent annually in 1970, had fallen to 1.64 percent in 1975. The 1976 world population of 4 billion, which had been projected to grow to 6.3 billion by the year 2000, now is projected to reach only 5.4 billion. These results were attributed to the success of birth control programs in the developing nations; in particular, China's birth rate, 35.5 per 1000 of population in 1964, had fallen to 14.0 by 1975. While subject to subsequent revision, these findings constitute a major piece of new evidence tending to support the optimistic views of this book.

ing nations to bring agriculture to something like that level; but to do so will not be easy. When one compares the combines and agricultural extension agents of Iowa to the bullocks and night-soil gatherers of China, the room for improvement is evident. It is not likely that there will be massive widespread famines which will depopulate whole countries. However, there will continue to be temporary local or regional food shortages. These will no more control the long-term future of humanity than the famines of pre-revolutionary China influence the present situation, where China is nearly self-sufficient in food. But such shortages will involve considerable human suffering.

The picture which emerges is of a developing world in which population is not a problem, together with an underdeveloped world in which large population increases will take place before control is finally achieved. However, in a world of independent nations, each country tends to reap the advantage and experience the disadvantages of whatever population policy it adopts. While the decades ahead may see famine and hunger, these will spur internal reform in the affected nations far more certainly than they will lead to the destruction of the world as we know it.

But population control is only one of the requirements for a worthwhile human future. There must be enough resources to maintain a high level of industrial activity. Neo-Malthusians argue that the resources of Earth are finite and will soon be exhausted, thus leading to a collapse of industrial civilizations. The alternate viewpoint is that most essential raw materials are practically inexhaustible in supply; that as we exhaust one raw material we can turn to lower-grade substitutes; and that eventually society can function using only renewable resources and elements such as iron and aluminum, which are abundant in the earth's crust. This latter viewpoint appears to be the proper one, and the neo-Malthusians appear to have been misled by their penchant for lumping all resources together without regard to their importance, ultimate abundance, or substitutability.

According to Alvin Weinberg and H. E. Goeller, of Oak Ridge National Laboratories, the world use of nonrenewable resources in 1968 totaled some 18 billion tons. Of this the percentages represented by individual resources were as follows:

Hydrocarbons, 66.60	Potassium, 0.07
Sand, 21.17	Manganese, 0.028
Limestone, 8.15	Copper, 0.022
Iron, 1.45	Zinc, 0.016
Nitrogen, 0.68	Silicon, 0.011
Oxygen, 0.45	Chromium, 0.007
Sodium, 0.45	Lead, 0.003
Chlorine, 0.45	Nickel, 0.002
Sulfur, 0.23	Titanium, 0.0002
Aluminum, 0.072	Tin, 0.0002
Phosphorus, 0.07	

Some of the metals are in short supply but can be replaced by substitutes. Electrical copper can be almost entirely replaced by aluminum; structural copper and brass are largely replaceable by steel, aluminum, or titanium. Chromium is used in making stainless steel, which for most uses can be replaced by titanium. Lead, principally used in pipes, can be replaced by plastic or plastic-bonded steel; the same is true for zinc (galvanized iron) and tin (tin cans).* In any case the minor metals are so small a part of the resource picture that they could increase in price manyfold and the overall economy would easily absorb the cost.

It is necessary to consider what present or prospective resources exist for the most extensively used materials and how long they may last. An indication of the latter is the ratio of the total resource to its 1968 rate of use, shown in the accompanying table.

Element	Resource	Maximum amount in best resource (%)	World resource (tons)	Resource-to-demand ratio in 1968 (years)
Hydrocarbons	Coal, oil, gas	75	1×10^{13}	2,500
Carbon	Limestone	12	2×10^{15}	4×10^6
Silicon	Sand, sandstone	45	1.3×10^{16}	5×10^6
Calcium	Limestone	40	5×10^{15}	4×10^6
Hydrogen	Water	11	1.7×10^{17}	$\sim 10^{10}$
Iron	Basalt, laterite	10	1.8×10^{15}	4.5×10^6
Nitrogen	Air	80	4.5×10^{15}	1×10^8
Sodium	Rock salt, seawater	39	1.6×10^{16}	3×10^8
Oxygen	Air	20	1.1×10^{15}	3.5×10^7
Sulfur	Gypsum, seawater	23	1.1×10^{15}	3×10^7
Chlorine	Rock salt, seawater	61	2.9×10^{16}	4×10^8
Phosphorus	Phosphate rock	14	1.6×10^{10}	1,300
Potassium	Sylvite, seawater	52	5.7×10^{14}	4×10^7
Aluminum	Clay (kaolin)	21	1.7×10^{15}	2×10^8
Magnesium	Seawater	0.012	2×10^{15}	4×10^8
Manganese	Sea floor nodules	30	1×10^{11}	13,000
Argon	Air	1	5×10^{13}	2×10^8
Bromine	Seawater	0.0006	1×10^{14}	6×10^8
Nickel	Peridotite	0.2	6×10^{11}	1.4×10^6
Titanium	Ilmenite, titanium-rich soils	32	2×10^{14}	9×10^7

The surprising and significant conclusion is that with three exceptions all of the most extensively used elements are available, in reasonable concentration, from resources which at 1968 rates of use would last for millions of years. Moreover, in extracting metals from these

---

*It is true that plastics are currently made from scarce oil or natural gas. However, plastics are so valuable that in the future it will be profitable to produce them from shale oil, deeply buried coal, or even carbon extracted from limestone. That is, sources of carbon or hydrocarbon, too costly to exploit for energy production, will not be too costly to exploit as raw materials for use in making plastics.

resources, aluminum from clay requires only 1.28 times as much energy as from bauxite; iron from laterites requires only twice the energy as from high-grade ores. Society can turn to these resources with little or no loss of living standard and would be based largely on glass, plastic, wood, cement, iron, aluminum, and magnesium.

One major exception is phosphorus from phosphate rocks, used for agriculture. Though known and potential high-grade resources are very large, they are hardly inexhaustible. In the long run, as H. G. Wells pointed out, we will have to recycle bones as fertilizer. In addition, we may have to extract some of the 0.1 percent of phosphorus available in ordinary rock. Agriculture also requires trace elements, which are slowly depleted by modern agricultural methods. Among these are copper, zinc, and cobalt, whose availability from nonrenewable resources is limited. So in addition to recycling bones, we will have to return other agricultural and animal wastes to the soil.

This hopeful resource picture is clouded, in the near term, by the most important exception: energy-producing hydrocarbons. For use in plastics, there is plenty of carbon in limestone; in addition, the topmost kilometer of shale in the earth's crust contains 200 times as much hydrocarbons as in coal, oil, and gas. Most of this is too dilute, however, to be used as an energy resource. In considering energy-producing hydrocarbons, it is found they constitute one of the scarcest nonrenewable resources: twenty to twenty-five parts per million of the top kilometer of Earth's crust. Yet they constitute two-thirds of the world's demand for nonrenewables. Moreover, they serve for more than energy alone. Coal is used in producing iron (though it is replaceable by electrolytic hydrogen) and electricity—mostly fossil-fuel generated—is needed to produce aluminum. Energy is also needed to smelt other metals or to produce substitutes for materials in short supply. So the only important resource shortage is one of energy-producing hydrocarbons. Our social and economic structures are unlikely to be disrupted because we have to use lower-grade ores and resources, provided that we find an inexhaustible source of cheap energy to substitute for hydrocarbons.

There are three energy sources which could prove suitable: fusion, fast-breeder reactors, and solar energy. All three are being extensively studied and developed by a number of countries.

Fusion power has been a popular option for over 20 years. Its advocates have painted glowing pictures of limitless energy from the deuterium of the sea, free of pollution or of radioactivity, producing only the clean ash of helium. This is the dream but the reality is much different. To begin, fusion differs from other proposed energy sources because it alone has been identified as an energy option before it has shown the ability to produce energy. Current fusion experiments produce something like one watt of power for every million watts fed in to run them. Needless to say, such power production is extremely costly. Moreover, power from fusion presents formidable technical problems which are not likely to be solved in the near future.

A number of experimental breeder reactors have been built and have produced power.

The breeder uses uranium-238, the common isotope which is 140 times as abundant as U-235, used in bombs. In the breeder, the U-238, which is not fissionable, absorbs a neutron and forms plutonium-239, which is, and which can be used to produce power. It also can be used to make nuclear bombs; it takes about 10 pounds of the stuff to make one. A large breeder reactor would have about 2000 pounds, at a value of $5000 per pound—three times higher than gold. By the year 2000, an energy economy based on the breeder would be producing 80 tons per year of plutonium-239 and at up to 500 shipments per week, the anticipated traffic, there would be ample opportunity for hijacking or diversion to the black market. If the breeder reactor is used as the means to bring the whole world up to American or European living standards, the world would need 15,000 reactors with old ones being retired and new ones being built at the rate of about 10 per week.

So fusion faces intractable technical problems; breeders face problems of safety. In addition, most proposed new energy sources have to meet economic goals. Currently, nuclear power plants and coal- or oil-fired generating plants are the most common means of generating electricity. All three represent predictable well-understood technologies; they generate power at costs in the rather narrow range of 1.8 to 3.0 cents per kilowatt-hour.

Cost estimates have been developed for several alternative energy sources. Geothermal power, which is already available in small quantities, looks to be in the range of 2.5 to 5.5 cents per kilowatt-hour. The fast breeder is in the range 3.5 to 5.5, while the best method of producing energy from fusion may be at 4 to 6. By far the most costly is ground-based solar power. The diffuse nature of solar energy, the need to track the sun in the sky, and the difficulty of building truly large single units, all combine to make the cost of a solar installation triple that of a breeder. The cost of electricity will be at least 7 and perhaps as much as 20 cents per kilowatt hour.

So the overall picture looks like this: To build a long-term livable world, we need population control in the underdeveloped countries, achievement of which will be aided by, and contribute to, a rising standard of living. Rising living standards demand continued supplies of resources, the availability of which is closely linked to the long-term supply of energy. And when we inquire into new energy sources, we find that they appear too costly, too unsafe, or too uncertain in their practicality to give genuine assurance that the world we seek is indeed achievable.

It is at this point that space colonization enters the picture.

The power satellites built in a space colony offer considerable promise as being the much-sought cheap, clean, and inexhaustible source of power. In space they can be built rapidly and at little cost to Earth. On the ground rectennas can be built at a cost of $1 or $2 billion for each 10-million kilowatt installation—a cost considerably lower than that of any ordinary power plant. The space program has often been criticized as expensive, a costly misuse of federal dollars. In fact, it is one of the least costly of the major United States programs, totaling about 1 percent of the federal budget. This low cost carries over into the use

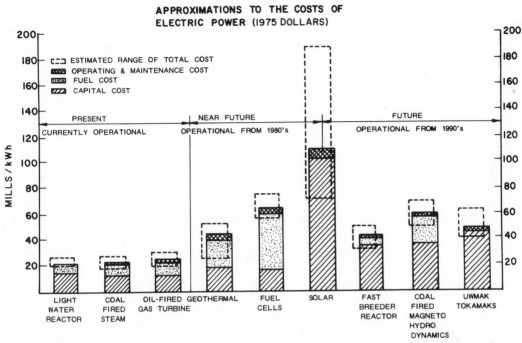

Comparative costs of present and future energy systems. All are more costly than power from power satellites built in space colonies, for which the cost is 3.5 to 8 mills per kilowatt-hour. (Courtesy University of Wisconsin)

of an expanded space program to solve the need for energy. In the power industry, 100 billion dollars is only a small percentage of what will be needed to meet the growing demand for electricity in the remaining years of this century. In space colonization, 100 billion dollars buys a complete space colony with its supporting lunar base and industrial facilities, ready to turn out powersats as desired.

The low cost of powersat electricity—as low as 0.35 cent per kilowatt-hour—means that it could be used to produce cheap hydrogen, which can be piped and burned like natural gas, by electrolysis of seawater. The hydrogen could also be used to make gasoline, with carbon taken from atmospheric carbon dioxide or from limestone and other carbonate rocks.

There have been numerous studies of the details involved in using liquid or gaseous hydrogen as a replacement for the portable fuels in use today. It is generally agreed that hydrogen, or synthetic fuels made with hydrogen, can replace today's fuels for whatever applications wished. The main problem is the large increase in electric generating capacity needed to accommodate its production. For this the solution is simple: build more rectennas, build more powersats.

The next century may see large numbers of rectennas serving as centers for major industrial parks. They will send some of their electricity by transmission lines to cities but much of the electricity will be used close at hand. There will be electrolyzer plants, producing hydrogen and also oxygen. There is no better way of treating municipal sewage or of removing pollution from rivers or lakes than with oxygen. Other plants will perform air separation, extracting atmospheric carbon dioxide and nitrogen. The former will be used together with some of the hydrogen to make gasolines, kerosenes, oils—a whole range of synthetic fuels and lubricants. Still other plants close to sources of energy and raw materials will turn out plastics and synthetic fabrics, fertilizers (using the nitrogen), pesticides, pharmaceuticals, and petrochemicals. Throughout the whole world: better things for better living through chemistry.

In principle there is a nice neat picture. Space colonies build powersats, which assure material prosperity and an abundant supply of resources, which promote the conditions for a stable population.

In reality, the picture is likely to be much more complex. The underdeveloped nations are not simply the Western nations minus electric toothbrushes. They have their own characteristic ways of living and thinking, which will change with time but now are no more relevant to such a future than the ecclesiastical feudalism of twelfth-century Europe is to the current advanced technological state of Western civilization. In many underdeveloped countries there is barely the foggiest notion of public trust. Corruption is a serious problem in the Third World but it need not be true indefinitely that the developing nations will be subject to the corrupt rule of cousins with black-market connections; these nations' desire for development can stimulate their ability to get things done.

Certainly, if they are to gain benefits from space colonization, they will demand arrangements which will assure them of their energy supply and not leave them subject to arbitrary price rises or to capricious interruptions. This should not be difficult to arrange. The rectenna, of course, will be located in the country it serves and under its control. Rectennas will be financed by the World Bank or by other international agencies rather than being subject to the uncertainties of foreign aid. The powersats themselves will be under the control of the colony, which will be responsible for their maintenance and upkeep and the colony's principal interest will be to grow, to build more powersats. The colony will wish to maintain friendly commercial relations with all nations. It will be very interested in proposals to lease the use of powersats or to sell their power at a fixed rate for a guaranteed term of years. After all, in their concern with growth and material development both the space colony and the countries of Africa or Asia will count as developing nations.

Beyond the material well-being, beyond the prerequisites for a decent world which will last a million years, lies the focus for human energy—the frontier.

Chapter 15

RING AROUND THE SUN

The moon glows softly in the night sky. Lovers admire it, poets sing songs to it, scientists contemplate its resources. But for a long-term source of materials for space colonization, it may be we will look, not to the moon, but to the asteroids. The asteroids are a collection of small bodies, located mostly between the orbits of Mars and Jupiter, with some crossing the orbit of Earth. They form a diverse population, rich in resources and in convenient size and location for human use for the centuries to come.

Today about 2000 are known well enough that they can be observed at will and a number of others have been observed sporadically. In all there are probably about 100,000 which could be seen with our largest telescopes. There is a continual gradation from large asteroids to small ones, to still smaller chunks of rock and down to the finest particles of dust.

However, they are not remnants of a shattered planet, nor are most of them in any way connected with meteorites. Superman's planet Krypton notwithstanding, planets don't just fragment or shatter, since they are held together by gravity. Asteroids today are regarded as material left over from the formation of the solar system, material which failed to coagulate into a planet. The asteroids are mostly in very stable orbits which stay far away from other planets; so chunks of rock which are knocked loose have no way to get to Earth.

Some common types of meteorites are also of great interest to the future of space colonization. There are iron and stony-iron meteorites, in many of which the metal is alloyed with nickel, ready for extraction by melting.

There are various types known as "chondrites," because they contain small inclusions known as "chondrules." These are more nearly similar to the primitive material from which planets were formed, in which iron was not separated out as if to make it ready for the rolling mill. Ordinary chondrites are about one-tenth iron; better still, they have up to 0.3 percent water. Even better in this respect are the meteorites of the most primitive composition, the "carbonaceous chondrites." These contain up to several percent carbon, which makes them very dark in color. Some of this carbon is in organic compounds; so carbonaceous chondrites typically have about half the organic content of oil shale. Best of all, they may have up to 20 percent water. They also have up to 0.04 percent nitrogen—not a lot, but worth having.

The metal content of these meteorites differs from the moon. They typically have only one-tenth the percentage of aluminum and less than 10 percent the titanium. But they are several times richer in magnesium and there is no lack of iron.

All these results are from laboratory studies on meteorites, and it is a big jump from the study of a 6-inch specimen 2 feet away to study of a 100-mile diameter asteroid, 100 million miles away. But through some rather startling advances in observational techniques, it is now possible for an astronomer to prospect for desired materials by working at the telescope, with better assurance of what he will find than if he were a terrestrial prospector with a mule and pickax—or, for that matter, with a spacecraft for multispectral photography of Earth.

Astronomers have long used the technique of measuring the apparent brightness of an object at two or three wavelengths and then comparing the measurements to obtain useful information. In 1968, Tom McCord of MIT devised a convenient means of measuring the apparent brightness not at 2 or 3, but at 24 wavelengths, giving a spectrum. He first used this instrument to study the asteroid Vesta, working with his associates John Adams and Torrence Johnson. To their surprise, repeated observations gave a spectrum which matched almost precisely a moon-rock spectrum which Johnson had on his desk. The surface of Vesta was made of the volcanic rock basalt—Vesta's surface had once been molten lava.

In 1969, an MIT graduate student, Clark Chapman, was looking for a topic for his dissertation. McCord suggested he survey the spectra of asteroids with his instrument, which he did. It took him two or three years because he had to use bits and pieces of time at the telescope, left for him by more senior astronomers. But he got both his data and his Ph.D. His data showed mainly that there were different types of spectra, few of which could be identified by comparison with meteorites or lunar samples.

Back in the laboratory Mike Gaffey at MIT did studies on the spectra of meteorite samples. In the meantime, other people were finding some very significant additional pieces to the puzzle. At Cornell, Joe Veverka was working on the problem of determining what fraction of sunlight is reflected by an asteroid, what astronomers call its albedo. He found a way to determine this by studying the degree to which the reflected light is polarized, as if it had passed through Polaroid glass. At Caltech, Dennis Matson was working on an entirely separate means of determining albedo from observations of asteroids in the infrared. One of the first asteroids he observed was Bamberga, number 324. He found its albedo was 0.03, meaning it reflects only 3 percent of the sunlight shining on it.

This meant Bamberga is one of the darkest objects in the solar system—darker than a blackboard, nearly as dark as soot from a fire. Only one thing could make Bamberga so black—carbon. Bamberga looked as if it might be a carbonaceous chondrite.

By 1972 it was possible to directly measure a number of important properties of asteroids and compare them with laboratory samples. By then, Clark Chapman had gone to Arizona, where it was much easier for him to get observing time on the big telescopes; Tom McCord stayed at MIT, along with Mike Gaffey. By mid-1974 these last two were able to write a paper for *Science* setting forth the basic rules for determining the composition of the surface of asteroids from the observed data, as well as giving the surface composition for fourteen asteroids. Ceres was found to be another carbonaceous chondrite, at least on the surface, as was Pallas. However, Juno looked like a stony-iron meteorite. The large asteroid Psyche, number 16, appeared to be a mass of nearly pure iron—not any iron, but quite likely the best grade of nickel steel.

Meanwhile Chapman and his associates were studying dozens of asteroids. At about the same time they announced an important general discovery: 90 percent of the main-belt asteroids fall into two categories. The first and larger group is similar to carbonaceous chondrites. The second class resembles stony-iron meteorites. Chapman called these classes C-type and S-type. But the asteroids whose orbits cross that of the earth or of Mars, were found in many cases to resemble a third type: ordinary chondrites.

All these studies represented work in astronomy by people who were entirely unaware of space colonization. It was Eric Drexler who pointed out the connection, in conversations with Mike Gaffey. If asteroidal resources were to be used to colonize space, there still remained a problem. As a source of raw materials the moon is close at hand and easy to reach. Most of the asteroids are several hundred times more distant. What's more, most asteroids' orbits are not nearly circular but are markedly elliptical, which makes them harder to reach. Worse, the orbits are tilted or inclined with respect to Earth's orbit, which makes reaching them even harder. This meant that mining the asteroids would be much more difficult than mining the moon, requiring far more advanced propulsion as well as extensive capabilities for supporting large crews for a long time in interplanetary flight. If space colonization using lunar resources looked like something that in a natural way could grow out of near-term efforts on lunar bases, large space stations, and space construction, then using asteroidal resources looked like something that would be several decades further off. After all, we had landed astronauts on the moon but have yet to send even a simple flyby mission to an asteroid.

Eugene Shoemaker came to the rescue with the discovery of 1976 AA. When Gene Shoemaker got his Ph.D in geology late in the 1940s, he decided that someday geologists would be able to study the moon at close quarters and that it would happen in his lifetime. He couldn't talk much about such wild notions with his colleagues; so he made the acquaintance of the few lunar astronomers, like Gerard Kuiper, who were engaged in what was then a lonely and neglected branch of astronomy. He kept at this all through the 1950s, much as Gerry O'Neill later would do with his private dreams of space colonies. Then came Sputnik and Shoemaker found he had other people to talk to.

It fell to Shoemaker to work out the major geologic ages or epochs in the history of the moon. On Earth in geologic history we have the Cretaceous and Permian and Silurian epochs. On the moon there are the Imbrian, the Procellarian, the Eratosthenian and Copernican—all named for major craters or maria.

In 1969 Shoemaker was picked as chairman of Caltech's Division of Geological and Planetary Sciences. But he found his administrative duties interfered with the work he loved; so he resigned the chairmanship after a couple of years and went back to doing research. What interested him now was not the moon but the asteroids.

After all, craters like those of the moon come about when a small asteroid, or a chunk of one, finds itself on an orbit which intersects that of Earth. He set up a program to search for these asteroids, with Eleanor "Glo" Helin doing the observing at Palomar. Their work led them to conclude that there should be 1000 to 2000 such asteroids, with diameters greater than a kilometer. This was a far larger number than previously proposed, making it appear that the "Earth-crossing asteroids" could be the long-sought principal source of the earth's meteors.

Then on January 7, 1976, "Glo" Helin observed a new asteroid about two miles in diameter. Since it was the first to be found in the new year, it was named 1976 AA. After it had been observed for several nights, there was enough data to compute its orbit, and to everyone's surprise, its orbit was closer to the earth's than that of any other known body except the moon. It circles the sun once every 347 days and for every 20 times Earth goes round, 1976 AA goes round 21 times. Its orbit ranges from 73 million to 106 million miles from the sun, compared to Earth's average of 93 million. In announcing the discovery, Shoemaker said: "Aside from the moon, this asteroid is one of the easiest places to get to in our solar system." When observed by telescope, it appeared to have the characteristics of an ordinary chondrite.

The only problem was its orbit, which was inclined to Earth's by 19 degrees. This would make it harder to reach than we would like. However, 1976 AA is only the thirty-fifth Earth-crosser discovered, out of the total population of 1000-2000; so there is opportunity to discover other small asteroids with orbits even closer to Earth's. What is wanted is an object with an orbit like that of 1976 AA, but with low inclination. Such an object could be reached with no more difficulty than is involved in going to the moon. Moreover, its small size would make landing on it like docking with another spacecraft. And while most such asteroids appear to be ordinary chondrites, there is no reason to doubt that some would be carbonaceous chondrites, which have more water and more carbon. The asteroid Betulia (number 1580), which nearly crosses Earth's orbit, is of this type. Nor are such asteroids small, since a one-kilometer diameter object has a mass of several billion tons.

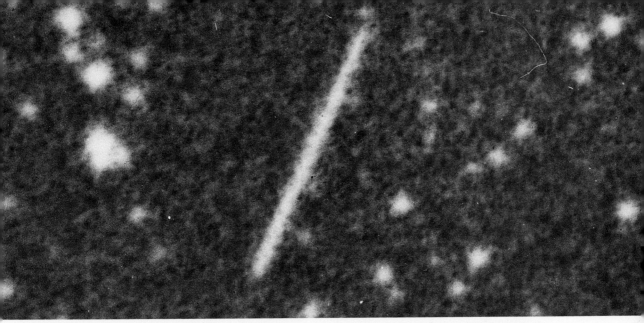

The discovery photo of asteroid 1976AA, which showed it as a streak on the photographic plate. Such streaks are all astronomers see of asteroids. (Courtesy Eugene Shoemaker)

This Viking photo of Mars' satellite, Phobos, shows a rocky, irregular, cratered, and pitted body which astronomers believe is similar to asteroids such as 1976AA. (Courtesy Jet Propulsion Laboratory)

The future of space colonization rests not only with specialists in propulsion and in other aerospace disciplines, it also rests with astronomers who will be searching for new Earth-crossing asteroids, which will be the first goals for expeditions beyond the moon. These will provide the much desired intermediate steps between manned lunar flight and manned flight deep into interplanetary space. The first steps will be only slightly more difficult than the flights which will then be in progress to the moon; yet they will begin the process of opening up the entire solar system to human settlement. As our astronauts prove out their techniques and develop their skills, they will push the space frontier deeper and deeper into the asteroids, until this frontier forms an ever widening ring around the sun.

At first, these small asteroids will be hauled and tugged into the vicinity of Earth to provide carbon and nitrogen and hydrogen for the growing colonies. For this the colonies will

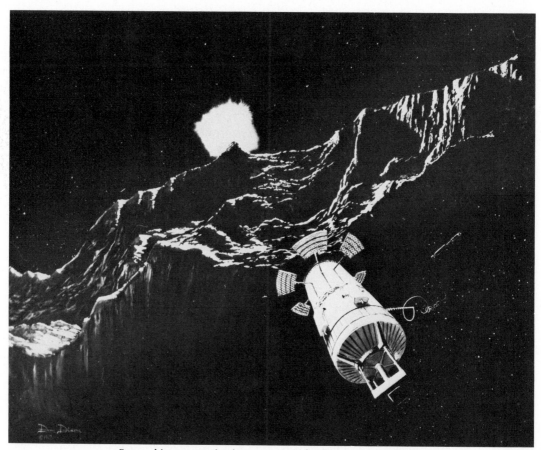

Spaceship reconnoitering an asteroid. (Courtesy Don Dixon)

build additional mass-drivers, similar to the one proposed for the moon, but designed for use as rocket motors to produce thrust. These will be maneuvered out to an asteroid of interest and used to move it. Or a chemical plant may be built, running on solar energy obtained with a large mirror, and sent to the asteroid. If the asteroid is a carbonaceous chondrite, a chemical plant can extract the organic material for shipment.

We may see the day when there are installations closely resembling oil refineries on the small C-type asteroids whose orbits are near that of the Earth. Carbonaceous chondrites typically have about half the organic-chemical content of oil shale. Most of this is a complex mixture of heavy hydrocarbons, including waxes or heavy oils as well as compounds related to benzene. The asteroids may contain a wealth of petroleum beyond the dreams of the richest Arabian sheik and one day it may be feasible to provide this to Earth.

However, the main activities at the refinery will be to extract water and to process the hydrocarbons into chemical fertilizers or plastics for use back at the colony. Nitrogen compounds will also be extracted and some of these will have a rather interesting nature, since they will be purines and pyramidines—nitrogenous bases such as adenine and guanine. These are the elements of the genetic code, when found in the DNA molecule, but they are also produced by nonbiologic processes and have been found in carbonaceous chondrites.

Some of the water may be electrolyzed to hydrogen and oxygen, providing rocket fuel for the shipment. Some of the plastics may be thinly coated with metal and formed into immense gossamer solar sails to use the pressure of the solar wind and sunlight for propulsion. The sail-borne ocean commerce of centuries past, the "argosies of magic sails" of Tennyson, may find their counterparts in fleets of solar sailships which undertake the long voyage to and from those distant outposts of commerce, visible only as streaks on a photographic plate.

In the long run, this will be seen as the first step toward removing from Earth much of the unpleasant or polluting industry and shifting it instead to an industrial base which is celestial rather than terrestrial. Nor will such a resource base face the problem of limits to growth. With the entire assemblage of asteroids available, limits are absurdly high: enough to support a human population some 20,000 times larger than exists today.

But in the short run—in the first few decades of space colonization—these resources will be useful mainly in stimulating two new trends in the growth of the colonies. There will be a trend toward building colonies of the largest size and another trend toward the evolution and growth of new or specialized communities, distinct from the main colonies.

The trend toward the largest colonies would carry out the initial plans put forth by Gerry O'Neill when he first started studying space colonies in 1969. The maximum size would be set by the strength of steel cables used in construction and the largest colonies would be strengthened with cables very much like those used in building suspension bridges. Today it is a matter for surprise to see how much larger these colonies would be than the Golden Gate or George Washington bridges. The time will come when the surprise and astonishment will be reversed. Future generations will be quite accustomed to the size of structures which can

Space colony of the largest size. Each cylinder is four miles across and over twenty miles long. Manufacturing facilities are on the forward ends, and rings of cylinders for agriculture surround each colony. (Courtesy NASA)

be built in space but they will regard with wonder and admiration the great bridges and skyscrapers of Earth, which had to be built in gravity.

The largest colonies would be twin cylinders, each four miles across and twenty miles long. They would rotate slowly in opposite directions once every two minutes to provide normal gravity. Being linked together, they will form a unit with no net spin, which can always be pointed toward the sun. The interior of the cylinders would be divided into long strips down their length: three immense window panels of glass, alternating with three strips of land.

There would be large mirrors outside to reflect sunlight through the window panels onto the land areas opposite. They would extend and retract to give night and day. Each cylinder could be surrounded by a ring of small units in which would be carried out the agriculture and manufacturing operations and there could be large solar mirrors and generating plants on the ends of the cylinders facing the sun. Inside these new worlds, clouds would form at the 3000-foot level in a blue sky—it takes 2 miles of atmosphere to make a sky blue. That atmosphere together with the soil in the land areas would be thick enough to protect against cosmic rays

Rainstorm inside a colony. (Courtesy Carolyn Henson)

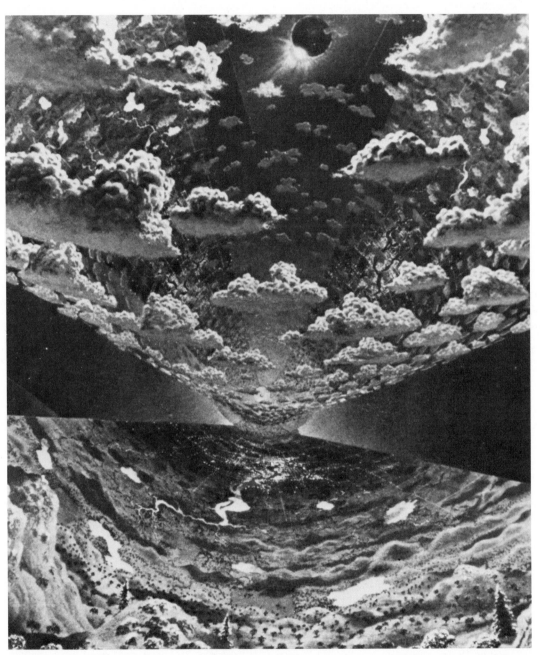

Solar eclipse inside a colony. The shadow of the earth is advancing up the valley; lights of towns are seen near the far end. (Donald E. Davis painting courtesy NASA)

Interior of one of the large colonies, showing emphasis on parks and lakes which is possible with such huge areas (hundreds of square miles) to work with. (Courtesy NASA)

while the sheer size of the colony would protect against danger from meteorites. The window areas would be divided into small panels by the strengthening cables, distant and barely visible, like suspension-bridge cables which appear as thin strands of silver when seen from several miles away. If a panel were broken by a meteorite, it would take 300 years for the air to leak away.

Such colonies, if populated to the density of the Stanford torus, would be home for 20 million people. But in fact there is no need for that many. The earliest colonies will be strongly tied to Earth and governed by considerations of economics. They will resemble the early English settlements in America, founded by gentlemen who were certainly excited by the prospects they saw—but who also expected to receive dividends from the East India Company. Those early space colonies will suffice to win our initial settlements in space and build enough power satellites to solve the energy problem. By the time they are ready to build the largest colonies, they will have the resource base and stable society which will make them largely independent of Earth. The large colonies will not be akin to the early English settlements. They will much more nearly resemble the American West, with its vast promise for those who would win its land.

As the large colonies proliferate, the early Stanford toruses will still represent valuable living space. But their interiors will be rebuilt to suit the open, well-forested styles in vogue in the middle of the next century. (Donald E. Davis painting courtesy NASA)

With cheap easy access to the resources of the moon and asteroids and space travel as routine as airline traffic is today, the largest colonies need not meet the same economic criteria as the Stanford torus. They can be designed as open, widely spread communities with comfort and beauty the main considerations. They may never be as open as the old West, with its half-dozen (or fewer) people per square mile, but their populations can easily be made one-hundredth as dense as the Stanford torus, approximating the layout of places like Columbia, Maryland, with its small communities of two-story single-family dwellings interspersed amid streams, forests and winding roads. The colonies' forests will not grow up overnight, even in the favorable conditions for agriculture possible in the colonies, but in time they will grow.

What will people do there? Some will be the agricultural chemists, the engineers and technical workers. They will be responsible for running the colony systems and for working with the production and manufacturing plants. Others will fill the hundreds of jobs which go to make up any community—storekeepers, plumbers, TV repairmen. With automated equipment to do most routine chores, few people will be unable to choose work which satisfies them. Many will be free to pursue occupations which are nonrepetitive and require a sense of art and beauty. In the overall space colonies there may be colonies of artists and writers.

Some of these people will form specialized communities and will develop (or bring with them from Earth) their own characteristic ideas of how life should be lived, how a community should be organized. On Earth it is difficult for these people to form new nations or regions for themselves. Indian reservations, the demands of some black people that they be granted title to several states of the Union, the founding of the state of Israel—at the cost of displacing the native Palestinians—are Earthly problems. But in space it will become easy for ethnic or religious groups, and for many others as well, to set up their own colonies. We may see the return of the Cherokee or Arapaho nation—not necessarily with a revival of the culture of prairie, horse, and buffalo, but in the founding of self-governing communities which reflect the distinctly Arapaho or Cherokee customs and attitudes toward man and nature.

Those who wish to found experimental communities, to try new social forms and practices, will have the opportunity to strike out into the wilderness and establish their ideals in cities in space. This in the long run will be one of the most valuable results from space colonization: the new social or cultural forms people will develop. It will even be possible for individual families, or small groups of families, to go off on their own and homestead an asteroid.

For this venture, the people would need a small spacecraft. They would also need equipment for agriculture and mining, all of which would cost perhaps $50,000 or $100,000—the price of a house in many parts of the country. The risks would be low and if they started out from the colony and their engine quit a few days out, they could stop and fix it, or radio for the Interplanetary Survey to pick them up. They could navigate with no more than a telescope, a sextant, and a small computer. It would be easier than navigating across the Atlantic, with its storms and surprises.

There may be millions of mini-asteroids, 100 feet or less in diameter, each one capable of supporting a small group. The people who will favor them will be like Daniel Boone, who said that when he could see the smoke from his neighbor's fire, it was time to be moving on. For rugged individualists, the Ayn Rand types who can't stand the restraints of organized society, these mini-asteroids may prove a haven. Like the lonesome cowboys of the last century, they may provide much of the romance and adventure of the high frontier.

All this is far from the usual science-fiction scenario of settlers bravely battling the hostile conditions of Mars or Titan, being strongly marked by the struggle in the process. We are not talking here about grafting extra pairs of lungs into the colonists so that they can live

Space colony founded by a group of expatriate San Franciscans. (Donald E. Davis painting courtesy NASA)

on Mars or equipping them with infrared vision so they can see more clearly in the gloom of Venus. Instead, we are talking about taking people as they are and providing them with worlds made to their measure.

Planets, after all, are poor places to live. They have high gravity, which makes it hard to get down to or off from their surfaces. It is as if ocean-going ships had to start their journeys by climbing out of a whirlpool. The gravity of planets also interferes with heavy construction and getting around. Planets usually have harsh weather, or weather which at the least is changeable and often unpleasant, and they are subject to extremes of temperature. Even if they are small and quiescent, like the moon, they still have day-night cycles which limit the usefulness of solar power, as well as mountains or valleys which as often as not are inconveniently located. In view of this and their own unparalleled opportunities by contrast, the colonists may say: "The trouble with people who live on planets is they think small."

Mars, the focus of so many hopeful dreams, might be bypassed. It will see its research centers for geology and other studies, but it appears to have few resources which cannot be had elsewhere. Even if it did, its gravity would make it costly to lift them out. Its atmosphere is just thick enough to prevent the use of a mass-driver. Yet the atmosphere is too thin to screen the solar ultraviolet or permit the use of aircraft for transportation. Mars of the great volcanoes, Mars of the deserts, of the frosty nights and the whistling winds in the canyons— if it is to be colonized, it will be done as an afterthought in the history of the human reach into space. It may remain a vast dry land, far from the major centers of commerce or population, thinly populated and of interest mainly to the people that live there. Mars may be the Australia of future centuries.

Chapter 16

COLONIZING THE STARS

In contemplating flight to the stars, the best understood fact is also the most important: the stars are far away. The closest are a million times farther than the closest planets, which in turn are about a million times farther than the distances that might be reasonably covered with a ten-speed bike. The comparison of a planetary mission to a ten-speed is not necessarily the measure of the technical advances needed to build suitable starships. There are no atomic bicycles but there have been plenty of proposals for atomic rockets and in designing a starship, the principal problem is to find a suitable means to apply nuclear energy to rocket propulsion.

To carry out interstellar missions in a reasonable time, such as a few decades, we need rockets with 1000 times the exhaust velocity of our current chemical ones, at least a few tens of millions of feet per second. The energy in the propellant has to be a million times greater than in today's chemical propellants and nuclear fuels have the happy property of possessing a million times more energy than chemical ones, which is why a one-megaton nuclear bomb weighs only one ton. But the problem of applying this energy to rocket propulsion has bedeviled more than one propulsion engineer.

There were proposals for nuclear particle emitters, the advocates of which believed they could produce 30 million feet per second of exhaust velocity. There were fission reaction particle emitters shaped into huge hemispheres, as well as radio-isotope sails of large area and thin, light construction. The propulsion specialist Robert Corliss has described their state-of-the-art as ''none'' and their usefulness as ''not a feasible propulsion system.''

In 1960 Robert Bussard proposed the interstellar ramjet. This idea involved a starship which would scoop up the incredibly thin, diffuse hydrogen of interstellar space and heat it and compress it to the conditions found in the center of the sun. It would undergo thermonuclear fusion, yielding helium, which would exhaust out the back. Bussard's concept had the advantage of being ingenious. It had the disadvantage of being wrong.

He assumed a drag-free starship, which in his context is a little like assuming a drag-free airplane. If such a plane could be built, it could fly at high speed just by burning atmospheric smog in its own ramjet. Such a smog-burner, of course, would give very weak thrust (smog is never *that* bad) and so would Bussard's ramjet. Hydrogen, at the conditions found in the solar interior, yields energy at the rate of two ergs per second per gram. A cooling bathtub releases energy a thousand times more rapidly. In terms of power-to-weight ratio, a Roman galley full of slaves would be more efficient. Even when the slaves were asleep you could get more energy from their body heat.

The topic of interstellar flight passed into fiction, such as "Star Trek." Too many engineering proposals had invoked too much hypothetical physics, sometimes together in anachronistic combination with current technology. It was as if Jules Verne had written of a steam dirigible carrying antigravity material for a lunar flight.

In 1971 Krafft Ehricke tried to predict the most advanced mission which might be undertaken with the technology then known. He chose the gas-core nuclear rocket for propulsion, a concept under study, which would contain a plasma of uranium inside a transparent wall, cooling the wall with the aid of a space radiator and achieving exhaust velocities up to 200,000 feet per second. He foresaw a multistage gas-core rocket, launched with the aid of gravitational tricks involving both Jupiter and the sun—and after 40 years reaching the distance of 0.1 light-year, one-fortieth of the way to Alpha Centauri.

At about the same time, Keith Boyer and J. D. Balcomb, leading nuclear propulsion specialists at Los Alamos Scientific Laboratory, offered the following assessment:

> *Specific energies of nuclear reactions are so high that system performance is limited by design constraints well before the specific-energy limit is reached. For example, the specific energy of uranium in fission [corresponds] to a maximum exhaust velocity of 40,800,000 feet per second, whereas the specific energy of deuterium plus tritium in fusion [corresponds] to a maximum exhaust velocity of 83,600,000 feet per second. Such exhaust velocities are only of theoretical interest. Many factors restrict any practical consideration to values up to 300,000 feet per second.* *

A few years earlier in 1965, James Strong had said much the same thing in his book *Flight to the Stars:*

> *The successful orbital flights of astronauts have tempted space travel enthusiasts to compare them with the pioneering days of heavier-than-air flight at the beginning of the*

*From American Institute of Aeronautics and Astronautics paper 71–636.

century. The analogy is misleading if it leads others to conclude that, sixty years from now, spaceliners will be crisscrossing the Solar System [as if they were] jet aircraft today. Men's first ventures into space would be better described as comparable to the hot-air balloon ascents in the eighteenth century. In effect, [some] 180 years must elapse [before] the luxury of fast interplanetary travel. Yet even by A.D. 2140, it would be surprising if a propulsion system for star flight were to exist. Another seventy years should see the broad outlines of a power unit emerge and take shape.

However, it did not take seventy years, but more like seven, for "the broad outlines to emerge and take shape"—not the twenty-third century but the year 1972. That year certain research findings were declassified, and for the first time it became possible to discuss star flight in terms of real feasible methods. At the same time it was found that there was intensive work under way, aimed at building what could only be described as laboratory experiments on the main thrust system for a starship.

This work is in controlled fusion. Since the invention of the laser early in the 1960s, it had been known that it was possible to take a pellet of liquefied thermonuclear fuel—deuterium and tritium, the heavy isotopes of hydrogen—hit it with a laser beam, and raise it to the temperature required for ignition, about 100 million degrees. This laser-zapping would not give a practical fusion reaction through merely heating the fuel, but it would do so if the laser were used to compress or implode the fuel to 10,000 times its normal liquid density. Although by the late 1960s laser fusion was being widely discussed, the most important concepts, including implosion, were made known only when the 1972 action of the Atomic Energy Commission declassified this line of research.

How can lasers compress a liquid? To do this while initiating fusion, it is necessary to deliver a pulse of laser energy to the pellet totaling some 10,000 joules—the energy of a small car at ten miles per hour. The energy must be delivered in the form of multiple beams focused to cover the entire surface of the pellet, which is less than a millimeter in diameter, and the delivery must take place in less than a nanosecond, a billionth of a second. This blows off the outer layers of the pellet at extremely high speeds. As the blown-off material accelerates, it generates a reaction force (by Newton's Third Law) that compresses the interior of the pellet. The system acts as a laser-driven spheroidal rocket whose payload is the rapidly contracting pellet interior. The imploding matter is accelerated inward to a velocity 50 times higher than Earth escape velocity. It collapses inward until internal pressure from its electrons brings the implosion to a halt. By then the pressure has been raised to a trillion atmospheres and the temperature and density are such that fusion is initiated.

Not all the mass of the pellet undergoes fusion. The fraction which does (and therefore the energy released) depends on the energy of the laser. Fortunately the fusion energy released increases more rapidly than the required ignition energy, so if the laser is powerful enough, the fusion process becomes efficient. To demonstrate laser fusion in the laboratory, there is under way currently a program of building a laser capable of delivering 10,000 joules to a pellet in 100 to 500 picoseconds (trillionths of a second) or with modifications, 50,000 joules. This facility at California's Lawrence Livermore Laboratory is to go into operation in 1977.

The most difficult problems involve the development of these large powerful lasers. For application to a starship or to the commercial generation of electricity, it will be necessary to have lasers with at least several hundred thousand joules in the pulse. For laboratory experiments, the big lasers are made from a well-understood material, neodymium glass. The problem is that all lasers are inefficient, requiring many times more energy to excite them than is subsequently delivered in the laser beam. For neodymium glass, the efficiency is about 0.1 percent. What is wanted is a gas laser with efficiency of at least 5 percent and work is now under way to find the best solution. Carbon-dioxide lasers appear to be particularly promising, offering efficiencies up to 10 percent, and are receiving nearly as much attention as neodymium glass. Other laser media which appear promising include hydrogen fluoride, argon, krypton, xenon, iodine, oxygen, and the vapors of the metals mercury, copper, or manganese.

It is too early to tell which of these media will be best but to aid in making the choice, Los Alamos scientists are working on a 100,000-joule gas laser facility which will be able to test these various media. To excite the lasers, to bring them to the condition to produce their pulses, considerable attention is being paid to the use of intense beams of high-energy electrons. These can excite large volumes of gas in a few nanoseconds. The electron beams themselves may serve to produce implosion and fusion of the pellets—a possibility which is attractive since electron-beam generators are more efficient than lasers.

As a result of all this, it now is possible to describe the broad outlines of a starship engine.

A useful starting point is the propellant tank, which contains a mixture of liquid deuterium and tritium. Fuel is pumped to a pellet-preparation system, where it is formed into a large number of small hollow spheres, resembling bubbles, each the size of a grain of sand. The pellets are frozen solid to aid in their handling and passed on to the injector.

The injector is a rapidly rotating device to accelerate pellets to about 10,000 feet per second and eject them into the engine thrust chamber. The high velocity allows a high frequency of fusion microexplosions, increasing the thrust of the engine. To do this, the injector is a small version of the rotary pellet launcher. It injects pellets at a rate of 500 per second.

Each pellet passes down a tube 300 feet long on its way to the thrust chamber. This allows its trajectory to be examined and corrected. The tube is equipped with optical sensors and lasers, which can be focused on one side or the other of the pellet. This blows off small quantities of its mass, producing a thrust which changes the pellet's direction of flight. In this manner the pellet is guided to the proper direction. At the same time its arrival time is predicted, so the laser can be primed to fire during the interval of 10 nanoseconds when the pellet will be passing the fusion point.

The thrust chamber consists of front and rear superconducting magnetic coils, mirrors, a pick-up coil, radiators, and structural tie bars. The superconducting coils produce strong magnetic fields to redirect the charged particles produced in the microexplosion, forcing them to stream out the back to produce thrust. These coils and other structural components are exposed to intense floods of X rays and neutrons. They can be protected from damage by having their exposed surfaces shielded with layers of beryllium and beryllium oxide. The coils and tie bars also must be cooled and for this there are layers of heat pipes inside the shields. These transport heat to the space radiators, where it is rejected at 2300°F. Insulation and recirculation of liquid helium then are enough to maintain the superconductivity of the coils.

The pellet enters the chamber and the laser fires. Its pulse is split into eight subpulses, which are directed onto eight mirrors mounted on the tie bars. The mirrors also are cooled with radiators. The subpulses converge on the pellet as it passes the fusion point, and about 25 percent of its mass undergoes fusion. About one-third of the energy released is in X rays and fast neutrons, the rest being in charged particles. These form a fusion fireball, a mass of intensely energetic plasma, which expands outward. The plasma soon encounters the magnetic field produced by the front and rear coils and pushes it outward, as if that field were a balloon into which air was being blown.

As it expands outward, the magnetic field is pushed past the pickup coil, which generates an induced voltage. The energy transferred to the pickup coil is about 5 percent of that generated in the microexplosion and it appears as electric energy. It is extracted from the pickup coil, transported down a transmission line, and dumped into an array of capacitors forward of the laser. The capacitors store enough energy to excite eight firings of the laser, so if one or a few pellets fail to undergo fusion, there still will be energy to fire the laser.

The plasma from the explosion is prevented from reaching any part of the thrust chamber structure by the magnetic fields. Instead, it is directed out through the rear coil. The thrust is 15,000 pounds; exhaust velocity is 30,000,000 feet per second, or 3 percent of the speed of light—a speed which is written as 3 psol.

The laser is a gas-filled cylinder 400 feet long by 20 feet in diameter. Passing through the central volume of the cylinder are heat pipes, which extend on either side of the cylinder to form the main starship radiator in the shape of two "wings." These run the length of the laser and operate at 2300°F, to maintain the laser at this temperature. Surrounding the laser are the elements of the electron-beam generator and its power supply fed from the capacitor bank. There is a small nuclear reactor to provide start-up power and to operate other onboard systems.

This is the starship engine. It is founded on well-understood principles of physics and it is essentially as simple as any chemical rocket motor. Quite likely within 10 years the problem of the laser will be well in hand, so we will know what gas to put in it. Then it will be a matter of engineering design and development, similar to that associated with any new type

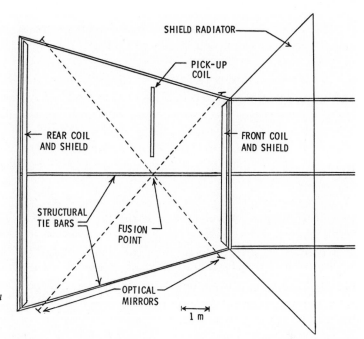

SHIELD RADIATOR

PICK-UP COIL

REAR COIL AND SHIELD

FRONT COIL AND SHIELD

STRUCTURAL TIE BARS

FUSION POINT

OPTICAL MIRRORS

1 m

Elements of a fusion-propulsion engine for a starship. (Courtesy Roderick Hyde)

of rocket. The result will be the highest-performing engine conceivable in terms of our understanding of physics. It will not be able to give speeds close to that of light, justifying the efforts of the platoons of mathematicians who have combined two trendy topics by studying the application of the theory of relativity to rocket flight. As it is, we will have to plod along at something like 10 psol—enough to get from Earth to the moon in 13 seconds. This means 43 years to the nearest star; a bit long, but manageable.

The longest planetary missions yet undertaken have involved sun-circling spacecraft such as Pioneer 6, launched in 1965 and still going strong. The longest ones seriously studied were a Jupiter-Saturn-Uranus-Neptune mission, in the Grand Tour program, with a duration of 13 years. The shortest interstellar missions could be only a few times longer than the longest planetary missions that are the stock-in-trade of JPL, the Jet Propulsion Laboratory. The initial interstellar missions, which will be unmanned, may be quite similar to the more advanced unmanned planetary missions which have been undertaken or may be flown in years to come.

Present-day missions will contribute to the stellar program particularly in the areas of computers, long-lived energy sources, and communications. For the initial missions, one key need will be an onboard computer which can be kept operational for several decades, while exhibiting the characteristics of artificial intelligence. Even today this problem appears well in hand. During its work on the Grand Tour program, JPL scientists were working on the

Fusion-driven starship with radiators glowing. (Courtesy Don Dixon)

STAR (self-testing and repair) computer, which would check itself out and replace failed units with spares when necessary. Such a capability itself requires a degree of artificial intelligence—the ability to recognize problems and exercise judgment on the basis of inputs from sensors.

JPL and the Stanford Research Institute are continuing to study artificial intelligence. For instance, on future Mars rovers it will be desirable to set down a tracked or wheeled vehicle which will take care of itself. If it comes to a crevasse it should not blindly fall in or, recognizing the danger, have to call up Earth: "Mission Control, there's nothing underneath the terrain-sensor stuck out in front, what should I do?" Instead, it should be able to recognize a crevasse when its TV cameras see one and maneuver around it. Well before the first starship is to fly, all these problems should be solved. Computer designers can do wonderful things with integrated circuits and can store unprecedented amounts of data in magnetic-bubble memories.

Unmanned star-probe, a successor to today's Mariner and Viking spacecraft, approaching a planet of a distant star. (Painting courtesy Don Dixon)

Communications will of course be interstellar in scope and will rely heavily on the techniques described in Chapter 13. One of the best reasons for building Project Cyclops will be to use it for communication with starships, just as the planetary program stimulated the building of conventional radio telescopes. A 100-meter paraboloidal antenna on the starship, fed with a megawatt of power and beaming a narrow-band signal at a frequency within the "water hole," will provide all the communicating power needed, and these systems have existed at least since 1960.

What instruments would such star probes carry? The large antenna would double as a radio telescope and would be a major instrument. The major item would be a large optical telescope carrying equipment to make observations in the ultraviolet and infrared, as well as at visible wavelengths. It would be equipped with the fullest range of instruments available to the modern astronomer: photometers, spectroscopes, radiometers, filter wheels for observation at selected wavelengths, polarimeters and image tubes, to name only a few. There would be arrangements to exchange lenses to permit a wide range of magnifications and fields of view. To permit even greater flexibility there would be a large mirror forward of the telescope to reflect the image of celestial bodies. Then the telescope itself need not swivel or turn—the mirror would be moved instead.

All this would be under the command of the starship computer. The computer would direct observations of the target star and other celestial objects of interest from the beginning of the flight. With months and years required for a signal from Earth to reach the ship, its instruments could do useful science only with a high degree of onboard autonomy. The computer would be primed to recognize anything new or interesting, to follow up the discoveries with more intensive observations using different instruments as appropriate, and to send back all the data over the radio link.

In the last five or ten years before arrival at its destination, the star probe's attention will increasingly turn to observing its target star and the near vicinity. At a distance of about a light-year, the star will be resolved as a disk rather than as a point of light. At about the same distance under highest magnification, the near vicinity of the star could begin to show luminous points which would be identified as planets. With 0.1 light-year to go, or one year to arrival, the largest planets would also show themselves as disks and the smaller planets or largest satellites would be evident. With a month to arrival, planets the size of the earth would show disks. At 10 psol, the starship would spend a full twenty-four hours in the innermost billion miles of the star system with the mirror forward of the telescope busily scanning the planets and satellites previously discovered and the data storage system soaking up the findings which guide the computer in its decisions and which subsequently will be relayed time and again to Earth, so that there is no danger of data loss.

This brings up the interesting question of how the missions should be conducted. An immediate possibility is to equip the ship with detachable spacecraft—orbiters, landers, sample returners; to provide it with extra propellant so it can slow down and maneuver at the destination and to let it fly from planet to planet (if it finds any!), visiting a new one every month or so while dropping off these spacecraft as its calling cards. Such an approach would certainly return vast stores of information and would give us a picture of the nearby stars and their planets more intimate than the one we now have of our own solar system.

The problem is that such rendezvous-class missions would need either more fuel or more time than the alternative, the fast-flyby missions which would send the star probes barreling through full tilt. With a desired mission velocity of 10 psol and a rocket exhaust velocity of 3 psol, it will take twenty times the starship mass in propellant. If we wish to speed up to 10 psol and then slow to zero at the destination, it will take 400 times the starship mass or if the propellant mass is to be kept at 20 times that of the starship, the mission time will be doubled.

At JPL it is a policy when visiting a new planet for the first time that the mission will be a flyby. Perhaps that policy will extend as well to the first star probes; however, hopefully at least a few rendezvous-class missions will fly, even accepting the longer flight times, to some of the nearer or more interesting star systems.

With such excellent onboard equipment, even the fast flybys will return data characterizing the star systems better than we now know our own solar system from telescopic observations only. It will be possible to detect and track planets and their satellites, determining their diameters. By following their motions, their masses as well as the details of their orbits will be found. Their temperatures will be measured and mapped at the telescope. Spectroscopic measurements will give the densities and compositions of their atmospheres, including particularly the presence of water or oxygen.

In the hours of closest approach color photography will be possible, showing surface features as small as 30 miles across. Major storms or other atmospheric features will be visible in detail as will many features of the surface. Polar caps, craters, volcanoes, rift valleys, mountains will all be seen—as will seas, continents and islands, lakes and river valleys, if they are there. Even if a planet is shrouded in cloud, it will be possible to study the surface using radar with the onboard antenna. And the methods described in chapter 15, used for determining the surface composition of asteroids, will serve also for these new applications.

Such starships will face hazards caused by the space environment. At 10 psol each molecule of interstellar gas will be a cosmic ray and particles of interstellar dust, a micron or so across, will strike with the intensity of sand blown in a tornado. Telescope lenses will need to be protected with glass plates, which can be replaced automatically as they become scratched or pitted. Fortunately our present knowledge indicates that meteoroids most probably do not exist in interstellar space (will this assurance someday constitute famous last words?), but when passing through a star system, the spacecraft may encounter particles the size of a grain of sand. These will strike with the energy of a ton of TNT.

For protection from such impacts, the star probe will need a self-propelled shield vehicle to ride out ahead. It will probably be thirty feet across and ride six miles ahead, acting as an umbrella against the rain of dangerous particles. It will have to be heavily built to withstand the impacts and with onboard propulsion and navigation so it can return to its station after an impact. The shield vehicle will be one of the more critical items of the star probe, which leads to the question: what shields the shield vehicle? The answer will probably be that the probe carries several such items, which will give continuing protection if one is lost. The large standoff distance means these vehicles will not interfere with astronomical observations, and for some purposes they may even be helpful. By blocking the light from a star, they can make it much easier to look for its planets.

With starships at hand, what stars shall we go to? The litany of the planets is familiar: Mercury, Venus, Earth, Mars, and on to Pluto. The roster of the nearby stars is not so familiar. There are a total of fifty-nine within twenty light-years, which has often been taken as representing a limit for exploration. If our eighteenth-century ancestors had spent their time building starships instead of fighting the French and Indian War, we would now have the data from most of the missions. The first few are: Alpha Centauri, Barnard's Star, Wolf 359, Luyten 726-8, Lalande 21185, Sirius, Ross 154, Ross 248, Epsilon Eridani, and Ross 128.

Of these Alpha Centauri is likely to be the focus of human hopes and expectations much as Mars has been. Mars is not only the closest planet, Venus excepted, it is the planet most like Earth. Alpha Centauri is the nearest star system; its two main components form a double star and both of them, by their luminosity and spectral characteristics, are found to be very like the sun.

There also are other stars similiar to the sun: Epsilon Eridani, Epsilon Indi, Tau Ceti, Omicron Eridani, 70 Ophiuchi, Eta Cassiopeiae, Sigma Draconis, 36 Ophiuchi, HR 7703, HR 5568, and Delta Pavonis. As is evident by their names, these are mostly the middle-of-the-road stars, not so brilliant as to be known by famous names, yet not so obscure as to merit only a designation from a star catalog.

In exploring these and others of interest, there will be recourse to a technique much appreciated at JPL. This is the technique of multiplanet flybys in which a spacecraft is sent past one planet on its way to the next. The Mariner Venus-Mercury mission of 1973–1975 was of this type, as is the 1973-1979 Pioneer 11 mission to Jupiter and Saturn. The years 1977 and 1978 will see the launching of two more missions of this type: another Jupiter-Saturn flight, and a three-planet mission to Jupiter, Saturn, and Uranus. In these flights the gravity of one planet changes the spacecraft direction and velocity to sling it on to the next.

Star-probe missions fly too rapidly to make use of the gravity even of stars to make course changes. But by reserving some propellant, it is possible for the starship to change its direction soon after passing the first star. It can visit two stars if the course change is not too large and can increase the scientific findings of a single mission.

A single flight can visit two very interesting stars: Barnard's Star, which may have planets, and 70 Ophiuchi, a triple star one of whose components is very like the sun. En route to the potentially significant Tau Ceti, a star probe could detour and visit Luyten 726-8, a double star whose components are faint and reddish, and one of which emits strong flares.

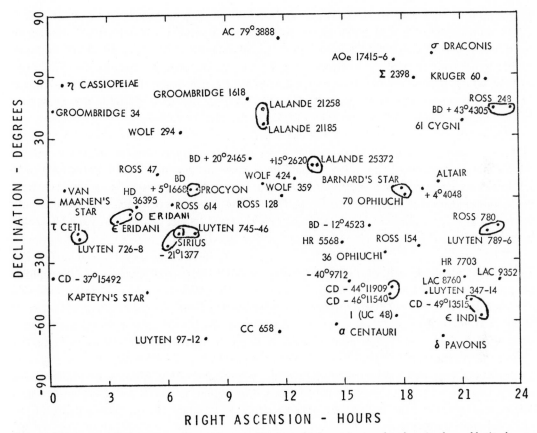

Stars within twenty light-years, plotted in coordinates which correspond to longitude and latitude on a Mercator map. Flyby missions to two stars in succession are possible to the star pairs circled. (Drawing by the author)

There could be a single mission to Epsilon Eridani and Omicron Eridani, and a flight to Epsilon Indi could then go on to the red dwarf CD − 49°13515. Sirius and Procyon would offer an additional inducement to visit them: their white-dwarf companions. These are highly compressed stellar remnants, in which something like the mass of the sun is contained in a volume only a few times that of Earth.

The ship would approach a white dwarf, gleaming balefully with its tiny disk. It would be pointed crosswise to its flight path to face the star and minimize heating and stress during the moments of closest approach. The white dwarf would loom up ahead, then flash past like some supernally brilliant light in a railway tunnel. In less than one second the ship would cross the whole of its surface and race on again into stellar space. The fierce gravity of the white dwarf would, in that brief encounter, so attract the starship that it would alter its

213

course up to several degrees. If the trajectory were off and the starship hit the white dwarf, in its last milliseconds it would make some very interesting observations which would never be recorded. It would strike with the energy of a trillion tons of TNT, or 10,000 of the most powerful nuclear bombs ever detonated. Yet so vast are the energies released by stars, even by faint white dwarfs, that such an explosion would be little more than a candle in a forest fire. If a companion star probe happened to be observing the white dwarf at that exact moment, it might—*might*—see a momentary flicker in the star's brightness.

These are the explorations and methods of discovery which will be carried into the next centuries. Such an ambitious long-term program of stellar exploration could only be carried out by the space colony. The colony would have both the resource base and the long-term stable future to carry it out and would have the industrial facilities to build the starships, those long and fragile creatures of zero g. The colony would recognize its long-term future prospects in the starships, for each night the colonists will face the stars, knowing them to be humanity's future home.

What sort of findings from the star probes would most encourage the colonists? The most interesting discovery, of course, would be a planet like Earth—green with chlorophyll, blue with oceans, white with clouds. Yet even if such a planet is someday found, the discovery would not end but begin the needed explorations. There would be the problem of its indigenous life forms. Would they be harmful to us, causing pestilences or diseases, perhaps? Or would we prove harmful to them? Would we compete with them? Would we perhaps find that so fruitful and hopeful a garden-spot, so rare an oasis, was the property of someone else? The galaxy is very old—life, even intelligent life, may prove rather common and humanity is very young. It is far from inconceivable that within a time much shorter than has elapsed since Jesus preached in Galilee, we may be called to account for our stewardship of the planet systems closest to us.

There is little prospect of realizing the science-fiction scene of fast interstellar ships carrying small parties of explorers. When humans go to the stars, they will go in large groups and they will be prepared to stay. But how can they be sure they will find a life-sustaining home?

To answer this, let us remember again the dangers of planetary chauvinism, of arguing that only on a planet's surface can people prosper. Let us remember the advantages of asteroids and other small celestial bodies in serving as a milieu for the growth of colonies. The answer is clear: in seeking to colonize the stars we should seek, not planets like Earth, but comets, small rocky bodies, including satellites of planets and asteroidlike debris left over from the origin of those stellar systems.

Collections of small bodies may be quite common. Alpha Centauri, a double star, may have no planets at all but each component may have extensive belts of asteroids which failed to form planets because of the gravitational influences of the other component. Collections such as these may be found and charted by even the earliest flyby missions, their surfaces studied and their compositions determined. The findings of the star probes can lead immediately to colonization efforts.

As the people of Earth reach stars and planets unknown, they will take with them much that is arcane or advanced in technology; yet they will also carry much that is simply human. (Artwork courtesy Don Dixon)

To colonize the stars, it will be necessary to build a small colony, along the lines of the Stanford torus, housing 1,000 people. Long experience with space colonies will have given our descendants an ample understanding of the problems in building closed-cycle systems in which there is neither leakage nor loss. To the colony will be fitted an array of starship engines together with an appropriate mass of propellant, surrounding it in an insulated structure to act as a radiation shield. With nuclear power for heat and light, the voyagers will live nearly as they and their forebears had always done ever since the dawn of space colonization.

If the target is Alpha Centauri or Barnard's Star, the youngest members of the community will still be alive at the destination. More commonly the voyage will far exceed a human lifetime. They will drift through space—these tiny bright spores of life, these islands of hope in the blackness. The distant stars, like lights seen across a valley in winter, will offer no aid. The voyagers will be on their own.

When their descendants see one star glowing more brightly than the rest, when it shows a disk and becomes evident as the target, when the ship slows and at length ends its journey hard by some nameless asteroid—then the colonizing of a new star system will begin.

Perhaps there will be a new Earth, a bright and blue habitable planet. Over the ensuing years, while the people work to recreate the colonies their ancestors had known, some will cautiously, tentatively approach it. They will send to it probes and orbiters, at length descending in small parties to its surface, and like Robinson Crusoe, perhaps they will find a footprint.

The star colonies will grow and prosper. They will communicate with Earth and each other, so star-flung humanity will not lose its common specieshood. In the thousands of years to come, new ships will set out to the star colonies. Nor will the people then forget their skills as spacefarers, for always the night sky will disclose further frontiers. There will be new starships, new voyages, originating not from our solar system but from other ones, and the history of the human reach outward will continue. In a few million years, in the time since the first pebble-tools, the whole of the galaxy may be the province of humanity. But long before then, we will quite likely have met other intelligent races far more advanced than we, and settled into our place in the community of interstellar civilizations.

We began with an introduction by Ray Bradbury. It is appropriate now to end with a quote from the same author, where he writes of the journey to the frontier in space:

> What of that journey, the rocket and its meaning, man and his endless ticketing of himself to Far Rockaway and Lands End and Copernicus Crater? Will we never get him away from the Viking longboat, off the trolley, out of the jet, free of the rocket or the damn time machine he so dearly wishes to invent, test, explode, and go far-traveling with?
>
> Never.
>
> Will any of it improve him?
>
> About as much as ten laps around a meadowfield and a cold shower help a boy of fifteen. It doesn't change him; it but makes him feel more alive.
>
> How can you possibly compare space travel with a sweaty boy and an icy shower?
>
> Because I want mankind to be very much alive. But improve him? No. Hitler and Stalin wanted to improve him out of existence.
>
> I would take him—warts, bumps, hogwash, mush, and all, every athlete's foot of him, armpit lumps, corns, bad dreams—and put him on the Moon, Mars, then drop him in the Coal Sack Nebula shouting with joy, shrieking with fear, and alive, alive, O!
>
> I don't think you can improve a thing that is already improved, already lost; always behind but always winning; filled with midnight, burning with sun; sly and untrusty, open and lacking guile.
>
> I sing paradoxical man.
>
> I accept not only his flesh but the bones within his flesh and the sin marrowing those bones.
>
> Approve of him? It is hard to approve of this lumpy child. But sons are always lovable, murderers though they be, saints though they be—and we hate saints sometimes, do we not, as much as we hate murderers?
>
> I sing the entire man, then, going into Space.*

---

*Excerpt from "Foreword" by Ray Bradbury in *Mars and the Mind of Man* by Ray Bradbury. Arthur C. Clarke. Bruce Murray. Carl Sagan. and Walter Sullivan. Copyright © 1973 by Harper & Row. Publishers. Inc. By permission of the publisher.

BIBLIOGRAPHY

The principal references, treating the subject of space colonization in a general manner and used in this book, are the following:

Space Colonization—A Design Study. Report of the NASA/ASEE 1975 Summer Faculty Fellow-ship Program in Engineering Systems Design. Eric Burgess, editor. NASA SP-413. Available from Dr. Richard Johnson, Mail Code 236-5, NASA Ames Research Center, Moffett Field, Calif. 94035. Wash-ington, D.C.: U.S. Government Printing Office.

Proceedings of the 1975 Princeton Conference on Space Manufacturing Facilities (Space Colonies). New York: American Institute of Aeronautics and Astronautics, in press. Available from Dr. Jerry Grey, AIAA, 1290 Avenue of the Americas, New York, N.Y. 10019.

"Lunar Utilization, Abstracts of Papers Presented at a Special Session of the Seventh Annual Lunar Science Conference." David R. Criswell, editor. Available from Dr. David R. Criswell, Lunar Science Institute, 3303 NASA Road 1, Houston Tex. 77058.

A collection of technical papers giving the results of the 1976 Summer Study is available as a NASA Report from: Dr. John Billingham, Bldg. 202, NASA Ames Research Center, Moffett Field, Calif. 94035. This collection will be subsequently published as a volume in the AIAA series, "Progress in Aeronautics and Astronautics."

The latest information about space colonies may be found in the *L-5 News,* the monthly magazine of the L-5 Society. This is the leading organization devoted to advocacy of space colonization. Sample copies are available upon request to Carolyn and Keith Henson, L-5 Society, 1620 North Park Avenue, Tucson, Ariz. 85719.

The following references are listed by the first chapter in which they were used. Many of them were referred to as well in preparing material for subsequent chapters.

Chapter 1

Dyson, F. J. "The Flow of Energy in the Universe." *Scientific American,* September 1971, pp. 50-59.

Hoyle, F. *Galaxies, Nuclei and Quasars.* New York: Harper & Row, 1965, pp. 132-160.

Hunten, D. M. *The Atmosphere of Titan.* NASA SP-340. Washington, D.C.: U.S. Government Printing Office, 1974.

Oliver, B. M., ed. *Project Cyclops.* NASA CR-114445. Moffett Field, Calif.: Ames Research Center, 1971, pp. 24-26.

Sagan, C. *The Cosmic Connection.* New York: Dell Publishing Co., 1973.

Sagan, C., and Shklovskii, I. *Intelligent Life in the Universe.* New York: Delta Books, 1966, pp. 246-257.

Wolfe, J. H. "Jupiter." *Scientific American,* September 1975, pp. 118-26.

Chapter 2

Brand, S. "Interviewing Gerard O'Neill." *CoEvolution Quarterly,* no. 7 (Fall 1975), pp. 20-28.

O'Neill, G. K. "The Colonization of Space." *Physics Today,* September 1974, pp. 32-40.

Chapter 3

Berman, P. A. "Photovoltaic Solar Array Technology Required for Three Wide-Scale Generating Systems for Terrestrial Applications: Rooftop, Solar Farm, and Satellite." Technical Report 32-1573, California Institute of Technology, October 15, 1972.

Brown, W. C. "The Receiving Antenna and Microwave Power Rectification." *Microwave Power* 5 (December 1970): 279-92.

Clarke, A. C. *Profiles of the Future.* New York: Bantam Books, 1967, pp. 12-21.

Ehricke, K. A. "Regional and Global Energy Transfer by Passive Power Relay Satellite." Space Division, Rockwell International Corporation, Report SD 73-SH-0017, April 1973.

Glaser, P. E. "Solar Power Via Satellite." *Astronautics and Aeronautics* 11 (August 1973): pp. 60-68.

Hammond, A. L. "Photovoltaic Cells: Direct Conversion of Solar Energy." *Science,* November 17, 1972, pp. 732-33.

"Microwave Power Transmission in the Satellite Solar Power Station System." Technical Report ER 72-4038, Raytheon Co., Jan. 27, 1972.

Miles, M. "Scientists Test Plan for Orbital Power Beams." *Los Angeles Times,* October 6, 1975, p. 3.

O'Neill, G. K. "Space Colonies and Energy Supply to the Earth." *Science,* December 5, 1975, pp. 943-47.

Williams, J. R. "Geosynchronous Satellite Solar Power." *Astronautics and Aeronautics* 13 (November 1975): pp. 46-52.

Chapter 4

Boyd, R. "World Dynamics: A Note." *Science,* August 11, 1972, pp. 516-19.

Brand, S. "Apocalypse Juggernaut, Goodbye?" *CoEvolution Quarterly,* no. 7 (Fall 1975), pp. 4-5.

Forrester, J. W. *World Dynamics.* Cambridge, Mass.: Wright-Allen Press, 1971.

Gillette, R. "The Limits to Growth: Hard Sell for a Computer View of Doomsday." *Science,* March 3, 1972, pp. 1088-92.

Heppenheimer, T. A., and Hopkins, M. "Initial Space Colonization: Concepts and R&D Aims." *Astronautics and Aeronautics* 14 (March 1976): pp. 58-64, 72.

O'Neill, G. K. "The High Frontier." *CoEvolution Quarterly,* no. 7 (Fall 1975), pp. 6-9.

Tuerpe, D. R. "A Two-Sector World Model." University of California Lawrence Livermore Laboratory, Preprint UCRL-75500, Livermore, California, December 1974.

Vajk, J. P. "The Impact of Space Colonization on World Dynamics." *Technological Forecasting and Social Change* 9 (1976).

Chapter 5

Barfield, C. E. "Technology Report/Intense debate, cost cutting precede White House decision to back shuttle." *National Journal,* August 12, 1972, pp. 1289-95.

Barfield, C. E. "Technology Report/NASA broadens defense of space shuttle to counter critics' attacks." *National Journal,* August 19, 1972, pp. 1323-32.

Castenholz, P. D. "Rocketdyne's Space Shuttle Main Engine." AIAA Paper 71-659, American Institute of Aeronautics and Astronautics, New York, N.Y., June 14, 1971.

Faget, M. "Space Shuttle: A New Configuration." *Astronautics and Aeronautics* 8 (January 1970): pp. 52-60.

Gillette, R. "Space Shuttle: Compromise Version Still Faces Opposition." *Science,* January 28, 1972, pp. 392-96.

Kline, R., and Nathan, C. A. "Overcoming Two Significant Hurdles to Space Power Generation: Transportation and Assembly." AIAA Paper 75-641, American Institute of Aeronautics and Astronautics, New York, N.Y.

Mueller, G. E. "The New Future for Manned Spacecraft Developments." *Astronautics and Aeronautics* 7 (March 1969): pp. 24-32.

1973 NASA Authorization—Part 2 (record of Congressional hearings). Washington, D.C.: U.S. Government Printing Office, 1972, pp. 720, 722-39.

"Shuttle Systems Evaluation and Selection." Vol 4, "Booster Data." Grumman Aerospace Corporation, December 15, 1971.

"Shuttle Systems Technical Review." Report SSV 73-26, Space Division, Rockwell International Corp., April 16, 1973.

"Special Report: Space Shuttle." *Aviation Week and Space Technology* 105 (November 8, 1976): pp. 9-151.

"System Concepts for STS Derived Heavy-Lift Launch Vehicles Study." Technical Proposal D180-18743-1, Boeing Aerospace Co., April 1975.

Chapter 6

Greeley, R., and Schultz, P., eds. *A Primer in Lunar Geology.* Moffett Field, Calif.: NASA Ames Research Center, 1974.

Heppenheimer, T. A. "Two New Propulsion Systems for Use in Space Colonization." *Journal of the British Interplanetary Society,* in press.

Heppenheimer, T. A., and Kaplan, D. "Guidance and Trajectory Considerations in Lunar Mass Transportation." *AIAA Journal,* in press.

Kolm, H. H., and Thornton, R. D. "Electromagnetic Flight." *Scientific American,* October 1973, pp. 17-25.

O'Neill, G. K. "Engineering a Space Manufacturing Center." *Astronautics and Aeronautics* 14 (October 1976): pp. 20-28, 36.

Chapter 7

Hagler, T. "Building Large Structures in Space." *Astronautics and Aeronautics* 14 (May 1976): pp. 56-61.

Ley, W. *Rockets, Missiles and Men in Space.* New York: Viking Press, 1968.

Schweikart, R. "Space colonies should keep away from the government for a while." *CoEvolution Quarterly,* no. 9 (Spring 1976), pp. 72-78.

Chapter 8

Farquhar, R. W. "The Utilization of Halo Orbits in Advanced Lunar Operations." Report X-551-70-449, NASA Goddard Space Flight Center, December 1970.

O'Neill, G. K. *CoEvolution Quarterly,* no. 9 (Spring 1976), pp. 50-51.

O'Neill, G. K. "Space Colonization and Energy Supply to the Earth." Testimony before the Subcommittee on Space Science and Applications, U.S. Congress, July 23, 1975.

Winter, D. L. "Man in Space: A Time for Perspective." *Astronautics and Aeronautics* 13 (October 1975): pp. 28-36.

Chapter 9

Bradfield, R. "Multiple Cropping: Hope for Hungry Asia." *Reader's Digest,* October 1972, pp. 217-21.

Todd, J. "The New Alchemists." *CoEvolution Quarterly,* no. 9 (Spring 1976), pp. 54-65.

Chapter 10

Carriero, J., and Mensch, S. "Building Blocks: Design Potentials and Constraints." Office of Regional Resources and Development, Cornell University, 1971.

Chapter 12

Hannah, E. C. "Radiation Protection for Space Colonies." *Journal of the British Interplanetary Society,* in press.

Wetherill, G. W. "Solar System Sources of Meteorites and Large Meteoroids." *Annual Review of Earth and Planetary Science* 2 (1974): 303-31.

Chapter 13

Black, D. C., and Suffolk, G. C. J. "Perturbing Aspects of the Companions of Barnard's Star." In *On the Origin of the Solar System,* edited by H. Reeves, pp. 336-38. Paris: Centre National de la Recherche Scientifique, 1972.

"Large Space Telescope—A New Tool for Science." Proceedings of a conference, January 30–February 1, 1974. American Institute of Aeronautics and Astronautics, New York.

Metz, W. D. "Astronomy: TV Cameras Are Replacing Photographic Plates." *Science,* March 31, 1972, pp. 1448-49.

Oliver, B. M. "The Search for Extraterrestrial Life." *Engineering and Science* 38 (December 1974–January 1975): pp. 7-11, 30-32.

"Space Erectable Large Aperture Reflectors." Report LMSC-A946613, Lockheed Missiles and Space Company, March 26, 1969.

"The Astronomer in the Engine Room." *Time,* October 27, 1975, pp. 75-76.

Van de Kamp, P. "Problems of Long-Focus Photographic Astronomy." *Applied Optics* 2 (January 1963): pp. 9-15.

———. "Alternate Dynamical Analysis of Barnard's Star." *Astronomical Journal* 74 (August 1969): pp. 757-59.

Wilcox, J. M. "Solar Structure and Terrestrial Weather." *Science,* May 21, 1976, pp. 745-48.

Wuenscher, H. F. "Manufacturing in Space." *Astronautics and Aeronautics* 10 (September 1972): pp. 42-54.

Chapter 14

Conn, R. W., and Kulcinski, G. L. "Fusion Reactor Design Studies." *Science,* August 20, 1976, pp. 630-33.

"Food and Agriculture." Special issue, *Scientific American,* September 1976.

Goeller, H. E., and Weinberg, A. M. "The Age of Substitutability." *Science,* February 20, 1976, pp. 683-89.

Metz, W. D. "Fusion Research (II): Detailed Reactor Studies Identify More Problems." *Science,* July 2, 1976, pp. 38-40.

Parnati, C., and Lord, M. "Population Implosion." *Newsweek,* December 6, 1976, p. 49.

"The Human Population." Special issue, *Scientific American,* September 1974.

Chapter 15

Chapman, C. R. "The Nature of Asteroids." *Scientific American,* January 1975, pp. 24-33.

McCord, T. B., and Gaffey, M. J. "Asteroids: Surface Composition from Reflection Spectroscopy." *Science,* October 25, 1974, pp. 352-55.

O'Leary, B. T. "Mining the Apollo and Amor Asteroids." *Science,* in press.

Chapter 16

Boyer, K., and Balcomb, J. D. "Systems Studies of Fusion Powered Propulsion Systems." AIAA Paper 71-636, American Institute of Aeronautics and Astronautics, June 1971.

Brandenburg, R. "A Survey of Interstellar Missions." Technical Memorandum M-30, Astro Sciences, IIT Research Institute, Chicago, July 1971.

Heppenheimer, T. A. "Some Advanced Applications of a 1-Million-Second I_{sp} Rocket Engine." *Journal of the British Interplanetary Society* 28 (1975): 175-81.

Hyde, R.; Wood, L.; and Nuckolls, J. "Prospects for Rocket Propulsion with Laser Initiated Fusion Microexplosions." AIAA Paper 72-1063, American Institute of Aeronautics and Astronautics, November 1972.

Martin, B. D. "Data Subsystems for 12-Year Missions." *Astronautics and Aeronautics* 8 (September 1970): pp. 55-61.

Nuckolls, J.; Wood, L.; Thiessen, A.; and Zimmerman, G. "Laser Compression of Matter to Super-High Densities I. Thermonuclear (CTR) Applications." *Nature* 239 (1972): 139-42.

Strong, J. G. *Flight to the Stars.* New York: Hart Publishing Co., 1965.

INDEX